SAFE STORAGE OF LABORATORY CHEMICALS

SAFE STORAGE OF LABORATORY CHEMICALS

Second Edition

Edited by

DAVID A. PIPITONE

A Wiley-Interscience Publication

JOHN WILEY & SONS, INC.

New York / Chichester / Brisbane / Toronto / Singapore

QD
51
S22
1991

In recognition of the importance of preserving what has
been written, it is a policy of John Wiley & Sons, Inc. to
have books of enduring value published in the United
States printed on acid-free paper, and we exert our best
efforts to that end.

Library of Congress Cataloging in Publication Data:
Safe storage of laboratory chemicals / edited by David A. Pipitone. —
   2nd ed.
      p.   cm.
   "A Wiley-Interscience publication."
   Includes index.
   ISBN 0-471-51581-7
   1. Chemical laboratories—Safety measures.   2. Chemicals—Storage—
   Safety measures.   I. Pipitone, David A. (David Andrew), 1952–
   QD51.S22   1991   90-12949
   542′.028′9—dc20   CIP

Printed in the United States of America

10 9 8 7 6 5

To my wonderful wife, Cheryl

CONTRIBUTORS

JOHN BEQUETTE, Scientific Materials Management Consultant, Columbia, Missouri

LESLIE BRETHERICK, Chemical Safety Matters, Dorset, England

FRANK L. CHLAD, F.L. Chlad & Associates, Aurora, Ohio

JACK GERLOVICH, Drake University, Des Moines, Iowa

E. LAMAR HOUSTON, University of Georgia, Athens, Georgia

ALLEN G. MACENSKI, TRW, Inc., Redondo Beach, California

L. JEWEL NICHOLLS, University of Illinois at Chicago, Chicago, Illinois

DAVID A. PIPITONE, Gallagher Bassett Services, Inc., Rolling Meadows, Illinois

PATRICIA ANN REDDEN, St. Peters College, Jersey City, New Jersey

NORMAN V. STEERE, Norman V. Steere & Associates, Inc., Minneapolis, Minnesota

FOREWORD
TO THE FIRST EDITION

The remarkably good record of the chemical industry in protecting its workers against on-the-job hazards is widely known. It is less widely appreciated that this reported low incidence of accidents is based largely on the favorable experience in our major chemical companies, where truly effective safety programs have been in operation for many years and company policies continue to strongly support good safety.

More chemical professionals work for small companies than for large, however, and in many of these small companies there has been a low degree of awareness of the potential hazards in handling chemicals. These companies have lacked safety specialists who are alert to dangerous situations and knowledgeable about preventive measures. As a result, accidents have occurred more frequently in the small companies (and these have not always been reported). Likewise, the safety practices in academic institutions generally have been far poorer than those in the well-run large chemical companies. Teachers have become increasingly aware of their deficiencies in the protection of students and staff, but they have felt helpless in the face of uncertainties about what should be done.

The papers that have been collected in this book are especially aimed at providing useful information for the newly safety conscious personnel of universities, colleges, and secondary schools and of small commercial operations. Concerns about the proper handling of chemicals in the stockroom and in bulk storage will be greatly relieved by using this book as a reference. Current and practical commentary is offered on relevant government regulations and on economic considerations essential for getting the most out of available dollars while ensuring compatibility with safety requirements.

These papers were delivered at a well-attended symposium in Kansas City sponsored by the American Chemical Society's Divisions of Small Business and of Chemical Health and Safety. The number of beneficiaries will now be greatly expanded by this publication. We are properly grateful to everyone involved in the project.

MALCOLM RENFREW
Professor of Chemistry, Emeritus

University of Idaho
Moscow, Idaho
January 1984

PREFACE

The late 1980s have seen a surge in the demand for chemical safety and information relating to a safe workplace, which will continue well into the 1990s. Federal regulatory activity gathered momentum, with encompassing requirements placed on organizations storing laboratory chemicals. The EPA's SARA regulations were published in 1986; hazard communication laws were broadened to include all workplaces in 1987; the EPA's chemical inventory reporting standards were published in 1988; the OSHA passed worker safety standards for hazardous waste and emergency response operations in 1989 and toughened the airborne contamination limits; and the Final Rule on the OSHA's Laboratory Standard was published in January 1990.

The upshot of all of this is that things change. I remember a high school chemistry teacher telling me of two one-pound bottles o.̈ sodium cyanide donated to the school storeroom. She hid them on top of a large cabinet in her chemical storeroom, for lack of a better way to store or dispose of the chemical. Eventually she retired from her teaching role. For all I know the sodium cyanide is still there, and from time to time, I wonder if the conditions or practices in that chemical storeroom have changed.

The storage of laboratory chemicals is not a static function. Storing chemicals safely plays a dynamic role in meeting the needs of an organization intending to "do chemistry." Change has prompted the second edition of this book—changes in legislation, changes in standards, changes in technology, changes as workers become more sophisticated.

The second edition contains both old and new material. The information and wisdom of some chapters were revised to reflect the dynamism of the past six years; and new chapters were written to highlight issues and provide a focus for an integrated approach to the safe storage of laboratory chemicals.

Each chapter plays an essential role in the total management of chemicals. Knowledge for the identification of chemical storage hazards is abundantly detailed. Practical solutions to minimize or eliminate the potential dangerous exposures in storage are discussed in depth. Management policies for achieving and maintaining safe chemical storage are outlined.

This edition follows the same pattern of organization as the first edition. The volume is divided into two parts: storage guidelines and case histories. Storage guidelines include an introduction to safe chemical storage; the impact of federal regulations on chemical storage; applications for flammables, corrosives, water-reactives, toxics, and other types of hazardous chemicals; engineering and safety standards; incompatible chemicals; the latest precautionary labeling procedures; emergency response and spill control; and the use of computers to manage and provide chemical health and safety information. Case histories include industrial laboratory examples, survey results, computerized inventory/warehousing for a large research concern, chemical storage at a major university, and further methods for getting unwanted dangerous chemicals out of schools.

The information in this book will be beneficial to storeroom operators, laboratory and safety directors, administrators, teachers, and consultants. Industrial hygienists and those who conduct safety inspections may apply this information to assess chemical storage areas. Finally, engineers and architects who design new storage and laboratory facilities (or who revitalize current facilities) will benefit from understanding the necessary safety principles.

My sincere thanks and appreciation for this second edition go to the individual chapter authors for their professional work and commitment to the field of chemical safety. I could not have completed this edition without their contributions. In addition, I would especially like to thank Joel and Cele Pipitone, Fred Potenza and Bill Stark of Gallagher Bassett Services, Jeffrey Stull of the Texas Research Institute, as well as all those friends and relatives who provided many good wishes and encouragement. I am indebted to Philip Manor of John Wiley & Sons, Inc., and Margery Carazzone for their unwavering support during editing and production. Saving the best for last, I cannot put into words how the care and concern of my wife, Cheryl, helped make this edition possible, and this volume is dedicated to her.

DAVID A. PIPITONE

Rolling Meadows, Illinois
November 1990

CONTENTS

PART 2   CASE HISTORIES

APPENDIXES

SAFE STORAGE OF LABORATORY CHEMICALS

PART 1

STORAGE GUIDELINES

CHAPTER 1

INTRODUCTION

DAVID A. PIPITONE
Senior Consultant
Gallagher Bassett Services, Inc.
Rolling Meadows, Illinois

The business of "doing chemistry" in laboratories centers around the activities of analysis, experimentation, research, and development. The past few years have seen an explosion in technological advancements, as witnessed by the rise of automated laboratory procedures, for example. The demand for laboratory services in industrial hygiene and environmental testing has increased dramatically. As the production of goods increases, so does the volume of laboratory tests. The research and development of new products, in particular chemical-related products, goes hand in hand with rapidly expanding fields such as electronics, telecommunications, and innovative materials research. Laboratory chemicals are the raw materials that support the delivery of laboratory services.

The hazards associated with the storage, handling, and use of laboratory chemicals have significant consequences. Personal injuries, damage to property and equipment, interruption of operations, and decreased laboratory efficiency have been the results of catastrophic events when chemicals are improperly stored, handled, or used. In recent years, federal and state regulations have been passed to reduce the risk of injury to workers and damage to the environment caused by the mishandling of chemicals. Insurance premiums for policies covering property damage, liability, and medical coverage for higher risk operations, such as laboratories, are on the rise.

The safe storage of laboratory chemicals is a vital concern for any laboratory. This chapter provides the rationale for and overview of chemical storage as a dynamic system. The concepts presented here are discussed and expanded in detail in the following chapters.

1.1 DEFINING SAFE STORAGE OF LABORATORY CHEMICALS

Webster's *New Twentieth Unabridged Dictionary* defines the verb *store* in the following ways: "(1) To stock to furnish; to supply; to equip. (2) To stock or supply with stores or provisions, etc., against a future time. (3) To deposit in a store or warehouse for safekeeping. (4) To store or lay up against future need; hoard." *Storage* is defined as "specifically, the act of depositing in a store or warehouse for safekeeping." These definitions imply both passive and dynamic components in the storage of materials. The passive component concerns the ongoing presence of an item "deposited" in a storeroom or a storage area—the stationary position until it is physically moved or removed. The dynamic component of storage involves the active process of the flow of goods into and out of their stationary positions. This component includes how items are *deposited* in the storage area and how they are *supplied* from the storage area.

These distinctions have important ramifications in the safe storage of laboratory chemicals. Both passive and dynamic components demand attention for each element and activity involved in the process of depositing and supplying laboratory chemicals. Many problems have occurred when chemicals are assumed to require no more care or attention than an inanimate object like a rug or a book when not in use. This is especially true when, once placed in their stationary positions in the storeroom, laboratory chemicals are assumed to be static and inert. Witness the accidents caused by peroxide formation, decomposition, and aging of chemicals. Attention must be given to the storage area's operations of supplying chemicals to laboratories. Dispensing reagents, repackaging bulk chemicals into daily usage containers, recycling solvents, and collecting hazardous waste activities often are accomplished in conjunction with the storage of laboratory chemicals.

Chemicals, especially laboratory chemicals, have a tremendous hazard potential entailing special requirements to avoid accidents. A safe storage process anticipates these requirements and provides the appropriate environment, procedures, and controls to prevent the release of hazardous energy from laboratory chemicals.

1.1.1 Storage As a Process

Storage of laboratory chemicals is a dynamic process that interacts with the overall cradle-to-grave system of chemical use within an organization. Figure 1.1 depicts some of these interactive relationships with an organization's overall functional process flow. Laboratory chemicals enter a facility through its receiving department and are moved to a permanent or assigned storage area. This storage area is the depository and the point of supply for laboratory use. Long-term depositing or storage of laboratory chemicals may take place in a centralized chemical storeroom or in designated storage areas. Short-term depositing takes place at the receiving and shipping dock and in hallways while the chemicals are en route to a more permanent destination. Midterm depositing often occurs in laboratories where reagents and other often used chemicals are kept close at hand for convenience. When not resting in a depository, chemicals are in transit. The handling (of closed containers) during transit is integral to the storage process. A balanced approach to the safe storage of chemicals considers this moving and depositing activity as part of the process flow of laboratory chemicals within an organization.

1.1.2 Achieving Safe Storage

The safe storage of chemicals does not just happen by itself. Like other facets of safety, a proactive approach must be employed to define requirements and measure compliance with those requirements. Requirements flow from the hazardous properties of the chemicals, established industry safety standards for those chemicals, and regulations passed by federal and state agencies, such as the Occupational Safety and Health Administration (OSHA), the Environmental Protection Agency (EPA), and other law-making agencies. Examples of requirements include performance-oriented standards for hazard communication, training of emergency response teams, requirements for personal protective equipment, and OSHA's general duty clause requiring an employer to provide a safe workplace.

To be effective, requirements must be translated into a process that evaluates risk and that plans, implements, and monitors activities and controls to achieve safety. Planning is the active process of devising a scheme to meet chemical storage requirements and to facilitate the ongoing storage process. Planning involves the who, what, when, where, and how analysis of chemical demand, design of facilities, determination of adequacy of controls, procedures development, and so on.

Implementation is the process of bringing the plan to life—providing controls, training personnel in procedures, and equipping the facility with protective equipment and supplies. Implementation shares both start-up and refinement modes. Start-up brings new facilities and operations on-line; refinement concerns the ongoing achievement of requirements in training new people and improving existing operations and facilities.

Monitoring the storage process involves reconciling actual storage conditions and operations with established chemical storage requirements. Monitoring is the feedback loop that identifies compliance with these requirements or points out the shortcomings in the storage process. Proactive monitoring gives rise to better plans and improved implementation. This cycle of planning, implementation, and monitoring is repetitive, and, if properly done, ultimately improves the safety process.

1.2 A SYSTEMS APPROACH TO SAFE STORAGE

Safe storage can be viewed as a system that undergirds the process of depositing and supplying laboratory chemicals. Figure 1.2 outlines some of the major elements and relationships necessary for such a system. The four main categories of system elements include (a) the environment in which the chemicals

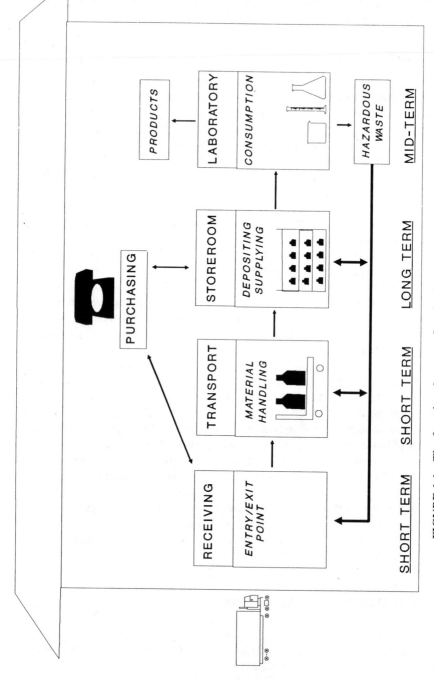

FIGURE 1.1 The functional process flow in the storage of laboratory chemicals.

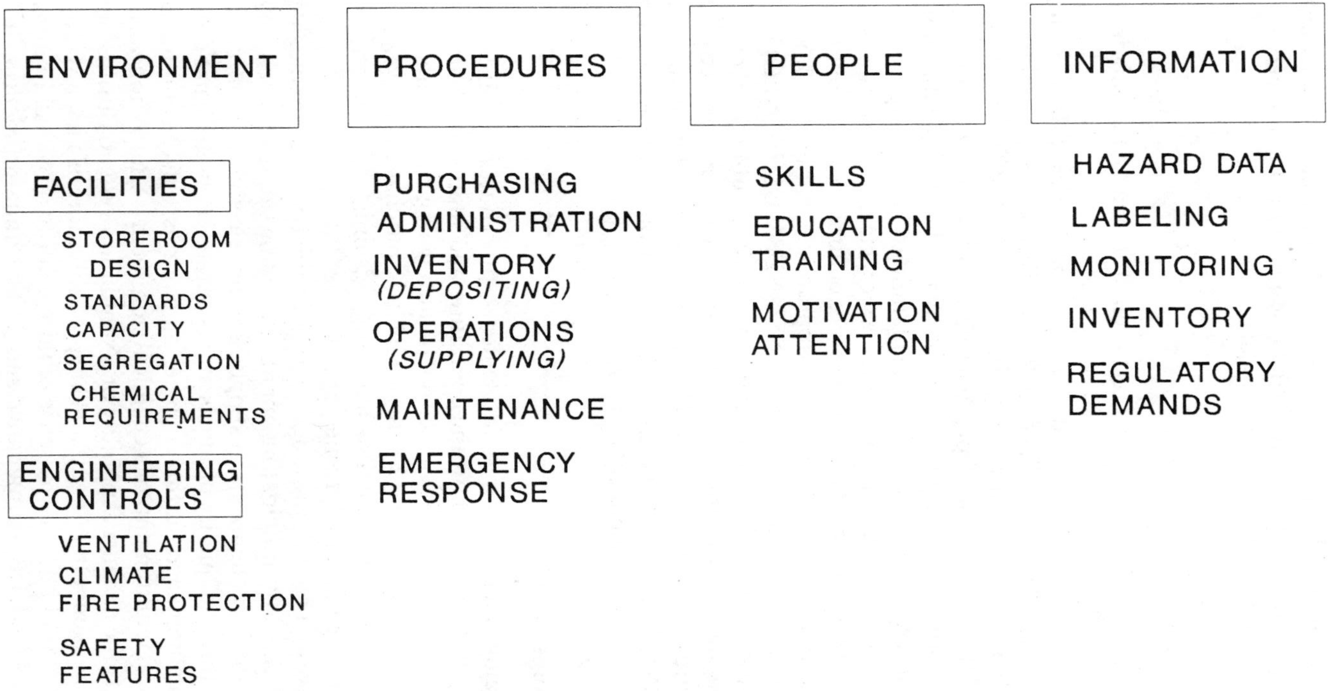

ENVIRONMENT

FACILITIES

STOREROOM
 DESIGN
 STANDARDS
 CAPACITY
 SEGREGATION
 CHEMICAL
 REQUIREMENTS

ENGINEERING CONTROLS

VENTILATION
CLIMATE
FIRE PROTECTION

SAFETY
FEATURES

PROCEDURES

PURCHASING
ADMINISTRATION

INVENTORY
(DEPOSITING)

OPERATIONS
(SUPPLYING)

MAINTENANCE

EMERGENCY
RESPONSE

PEOPLE

SKILLS

EDUCATION
TRAINING

MOTIVATION
ATTENTION

INFORMATION

HAZARD DATA

LABELING

MONITORING

INVENTORY

REGULATORY
DEMANDS

FIGURE 1.2   Elements necessary for a system of safe laboratory chemical storage.

are deposited and kept until future use, (b) the operations by which chemicals are deposited into and supplied from storage areas, (c) the people charged with storage duties and responsibilities, and (d) the acquisition and management of information about the laboratory chemicals.

1.2.1 Environment

The cornerstone of safe chemical storage is the environment of the long-term depository, the chemical storeroom. Planning a safe environment for chemical storage is most effective after analyzing the organization's demand for laboratory chemicals. The adage "design follows purpose" can be applied. The storage environment consists of facilities, equipment, physical provisions, and engineering controls to provide safe conditions for the long term, while accommodating the functionality to satisfy the demand of laboratory consumption. Storeroom design includes the planning of physical space, protective measures (fire protection, emergency water, etc.), safety equipment (cabinets for flammables, etc.), and engineering controls (ventilation, temperature and humidity control) to achieve appropriate storage conditions. The incorporation of national standards (ANSI, NFPA, OSHA, NEC) into design parameters is critical. Examining the nature of different types of chemicals to be deposited in the chemical storeroom reveals the need for specific safeguards, such as those for highly flammable, toxic, or reactive chemicals. Another design issue concerns making provisions for the segregation of incompatible classes of chemicals.

1.2.2 Procedures

The second major category of system elements, procedures, involves the activities and operations that allow the depositing and supply of laboratory chemicals to take place. These are carried out in the ongoing implementation/operations phase of safe storage. Examples include purchasing, preventive maintenance, inventorying, material handling, administrative functions, transactions, and the handling of emergencies.

Administrative and purchasing functions play an important role in the safe storage of chemicals—by anticipating but not exceeding storage capacity, by keeping a fresh stock of chemicals in supply, and by selecting vendors committed to safety through quality products and accurate safety information.

Preventive maintenance on the engineering controls, safety equipment, and other physical requirements of a chemical storeroom is vital. A nonworking blower limits the exhaust capability of a ventilation system. Burned-out batteries in an emergency lighting system are probably more of a hazard than no emergency lights at all—for the sense of false security provided. Routine inspections and scheduled maintenance should be standard practices in a chemical storeroom.

Proper material handling is important in the transport and supply of laboratory chemicals within an organization. This includes the necessary equipment, practices and procedures, written rules or guidelines, and security of containers to prevent accidental releases.

Inventorying laboratory chemicals, the process of depositing them in the storeroom, must follow the design of the storeroom to segregate incompatible chemicals; accommodate toxic, flammable, or reactive chemicals; prevent accidental releases; and provide a smooth retrieval and restocking process.

Accidents do occur, despite the best controls and prevention techniques. Preparing for emergencies is a good form of loss prevention. This includes having a written disaster plan, emergency procedures, appropriate safety equipment, personal protective equipment, and an available, trained response team. Controlling an emergency and mitigating its effects are necessary to reduce the severity of injuries or damage from the accident.

1.2.3 People

People are the third major element of a safe storage system. Accidents are caused by unsafe acts and unsafe conditions. The majority of accidents are caused by unsafe acts. People move the chemicals; people order the chemicals; people deposit the chemicals; people handle and use the chemicals; people respond to chemical emergencies. The players in the storage process are more than just the storeroom manager or attendants: consultants, architects, maintenance personnel, safety officers, the emergency response team, even fire fighters are among those who may be involved as a result of storage operations. In the case of mid-range chemical storage in laboratories, the activities of laboratory managers, chemists, and technicians affect how the chemicals are stored. The skills and training of persons in the safe storage of chemicals are influenced by how knowledgeable they are of the procedures and how informed they are about chemical hazards. Paying attention to detail and having the motivation to interpret and respond to warning signs are needed to anticipate and correct hazards before catastrophe strikes.

1.2.4 Information

The fourth major element in the safe storage system is the acquisition, communication, and management of information. This process involves obtaining ongoing data on the status of the stored chemicals, the environment, the operations, and the people involved. It includes obtaining, filing, and making available Material Safety Data Sheets (MSDSs) to comply with federal and state hazard communication legislation. It includes labeling chemical containers properly to warn of potential danger, injury, or reactivity. It includes educating and training personnel in the proper practices of safe storage. It includes monitoring the current condition and status of the environment and the chemicals themselves, such as for signs of decomposition, time-related formation of peroxides, and so on. It includes the reporting of inventory to the local fire department to comply with Superfund Amendment and Reauthorization Act (SARA) requirements. It includes obtaining, assessing, and complying with other regulations published by federal, state, and local agencies. Fortunately, managing information today has become easier through the use of computerized resources and software—databases, modeling programs, and user friendly retrieval interfaces.

1.2.5   Safe Storage: An Integral Part of Organizational Safety

The safe storage of laboratory chemicals is only a part of the formal safety program of an organization. As a component, safe storage should be integrated and managed with the overall safety program. Standards, rules, and communication channels must be planned, adapted, and developed to achieve the protection and preservation of personal health, optimal use of resources (especially chemical inventory), and continuous operations. Formal reporting relationships, responsibilities, and duties should be outlined and assigned to storeroom personnel, laboratory managers, shipping and receiving workers, and other persons involved in the chemical storage system.

The formality and sophistication of the safety program varies with the size, type, and philosophy of the organization. The process for planning, implementing, and monitoring chemical storage should be under the direction and responsibility of one person. A written policy statement, operating procedures, and storage requirements should be communicated to those with line responsibility for storing laboratory chemicals. It is important for the assigned safety officer to lead and direct the implementation of these safety initiatives. This direction and leadership should flow with the overall safety mission of the organization to create awareness for safety in chemical storage.

1.3   CONCLUSION

Every organization that uses laboratory chemicals needs an ongoing system for the safe storage of those chemicals. The complexity of the system and its major elements—environment, procedures, people, and information—depend on the type and amount of chemicals stored. The chapters included in this book address the issues involved in these system elements. The key to an effective and safe storage system is the analysis of your storage needs, both in depositing and supplying chemicals to your laboratories. As a system, safe storage is an ongoing process, a continuous cycle of planning, implementation, and monitoring strategies. Safe storage is not achieved by just buying safety cans or cabinets for flammable liquids. It requires regular attention and management. The safe storage of laboratory chemicals is an achievable goal that makes the workplace a better place to "do chemistry."

CHAPTER 2

THE IMPACT OF FEDERAL REGULATIONS ON THE STORAGE OF LABORATORY CHEMICALS

DAVID A. PIPITONE
Senior Consultant
Gallagher Bassett Services, Inc.
Rolling Meadows, Illinois

2.1 INTRODUCTION

Federal and state laws regulating the use of chemicals in the workplace have evolved at a rapid pace over the past 15 years. In the 1980s alone, hundreds of pages of new and proposed federal regulations have appeared in the *Federal Register*. The 1989 Code of Federal Regulations contains many pages of codified laws concerning chemicals.

The intent of these laws is to protect the health and safety of the worker, the community, and the environment. Federal and state regulations place both a legal liability and legal obligations on an organization. Failure to comply with the requirements of these regulations exposes an organization to the legal liability of fines, penalties, and even criminal prosecution. The legal obligations are fulfilled and freedom from legal liability is achieved when compliance with requirements is integrated into the day-to-day operations of an organization.

The purpose of this chapter is to identify and highlight implications and outline requirements of the federal regulations that have an impact on laboratory chemical storage. The system components for storing laboratory chemicals are subject to various federal regulations. In Chapter 1 the components of a chemical storage system were identified: environment, procedures, people, and information. Each component presents potential hazards that can impair the health and safety of workers, the community, and the environment if not adequately managed and controlled.

Employers and employees involved in chemical storage in the 50 United States and all territories fall under federal and state Occupational Health and

Safety Administration (OSHA) regulations. The reader should note that the states may adopt specific federal regulations or adapt and refine them into more stringent laws. The minimum requirement for state regulations is that they must be as strict as the federal regulations; there is no limit on more strict regulations. This chapter presents federal regulations appropriate to chemical storage operations that have been adopted or proposed through June 1990. A discussion of the specific requirements of state statutes is outside the scope of this chapter.

2.2   THE FEDERAL REGULATORY PROCESS

Law making is part of a dynamic, ongoing process that gives birth to initial and evolving regulations. At the federal level, bills are introduced in either the U.S. House of Representatives or the U.S. Senate. Each bill is numbered and submitted to a committee for consideration. The assigned committee(s) investigate and study the issues, holding hearings and soliciting comments. A report is prepared and issued with a recommendation of whether the bill should become law. Once issued, the bill is put on the respective congressional calendar, considered, debated, and, if passed, it becomes an act. If the act passes both the House and the Senate and is signed by the President, it becomes law.

Statutes, which are part of the law, empower an administrative agency, such as OSHA or the Environmental Protection Agency (EPA) to develop and publish regulations. Both the EPA and OSHA follow a similar rule-making process. An intention to develop proposed regulations is published in the *Federal Register* through Notices of Proposed Rule-Making. The agency publishes proposed standards and gathers information through written comments or public hearings concerning the proposed regulations. After the comment period is closed, the agency reviews the comments. Final regulations are formulated following a determination of their feasibility and cost impact on the "regulated community," and interaction with the Office of Management and Budget (OMB). Final regulations are published in the *Federal Register* with an effective date, when they will have the force of law.

2.3   OSHA REGULATIONS

The OSH Act of 1970 established OSHA, a federal rule-making body and administration, to regulate the safety and health of workers in organizations. OSHA regulations that involve the storage of laboratory chemicals are found in Title 29, Code of Federal Regulations; Part 1910 (29 CFR 1910), also known as "General Industry Standards." Many OSHA regulations have specific requirements for compliance based on either national consensus standards—especially those of the American National Standards Institute (ANSI), the

National Fire Protection Association (NFPA), and other voluntary standard groups—or other existing federal regulations, such as those issued by the Department of Transportation (DOT). A recent trend in the OSHA regulation-setting process has been the publication of performance-oriented standards that set a level of performance for organizations to comply with the intent of the law but leave the means to achieve the performance level to the organizations. The following sections outline OSHA regulations for the system components of the chemical storage process.

First, to help the reader understand cited references, a summary of how the OSHA regulations are organized is given. OSHA regulations are divided into subparts which contain one or more paragraphs. Subparts are designated with capital letters, such as Subpart Z, Toxic Substances. Each subpart has several paragraphs, which are numerical and prefaced with *1910*. For example, Subpart H, Hazardous Materials, has several paragraphs, one of which is 1910.106, Flammable and Combustible Materials. Each paragraph has subparagraphs, which are designated by lower case letters, for example, 1910.106(a). Subparagraphs are further organized. Each subpart contains a paragraph on the Sources of Standards (from which the regulations are developed), along with another paragraph identifying the Standards Organizations. Paragraphs normally contain definitions, the scope and application of the regulations set forth, and the specifications and requirements of the law.

2.3.1 Regulations for the Chemical Storage Environment

OSHA stipulates specific requirements for engineering controls and facility arrangements based on the hazard of the materials stored.

2.3.1.1 *Means of Egress*   Subpart E of 29 OSHA 1910 contains general and specific minimum requirements for exits and egress in cases of fire and other emergencies. Paragraph 1910.35 defines the terms in this subpart; Paragraph 1910.36 spells out general and fundamental requirements of exits; Paragraph 1910.37 specifies the requirements for means of egress. Chemical storage areas, especially storerooms, that contain exits or are along the path of emergency egress must comply with the provisions of these sections. Of particular note are those found in Subparagraph 1910.37(q), Exit Marking, especially Subparagraph 1910.37(q)(2), which requires that "doors to chemical storage areas, not leading to an exterior exit, but located along an egress path, . . . shall be identified as . . . Storeroom."

2.3.1.2 *Walking/Working Surfaces*   Subpart D of Section 1910 concerns itself with the safety of walking and working surfaces, which can be applied to chemical storage areas. Paragraph 1910.22 sets forth the general requirements for (a) housekeeping, especially for clean, orderly, and sanitary storerooms and clean, dry, and clear floors; (b) clear aisles and passageways, which have sufficiently safe clearance for material handling equipment (e.g., carts, drum trucks); and (c) provision for floor loading capacities.

In addition, Paragraphs 1910.25 (portable wood ladders) and 1910.26 (portable metal ladders) have implications for storage operations, to varying degrees. Where storage operations involve stairways, wall or floor openings (1910.23), fixed industrial stairs (1910.24), fixed ladders (1910.27), or scaffolding (1910.28), managers should determine the need for compliance with requirements in those sections.

2.3.1.3 Hazardous Materials

Subpart H of the OSHA standards covers the requirements for operations that involve hazardous materials. For storage operations of laboratory chemicals, this subpart includes general requirements for compressed gas cylinders (1910.101), acetylene (1910.102), hydrogen (1910.103), oxygen (1910.104), and flammable/combustible liquids (1910.106). [Also included in Subpart H are storage of anhydrous ammonia (1910.111), the use of spray operations with flammables (1910.107), the use of dip tanks with flammables (1910.108), requirements for explosives (1910.109) and large tanks of flammables (1910.110), which are not a normal part of common laboratory chemical storage operations. A discussion of these sections is outside the scope of this chapter.]

The minimum OSHA safety requirements for compressed gas cylinders, expressed in Paragraph 1910.101, include determining their safe condition by visual inspection, storage and handling in accordance with Compressed Gas Association Pamphlet P-1-1965, and the installation and maintenance safety relief devices on cylinders. The transfer, handling, and storage of acetylene cylinders must be in accordance with Compressed Gas Association Pamphlet G-1-1966.

Paragraph 1910.103 specifies the requirements for gaseous hydrogen systems in which the hydrogen is delivered, stored, and discharged in gaseous form to a facility's piping. These regulations would apply to those chemical storage operations that receive hydrogen from outside suppliers on mobile equipment.

Paragraph 1910.104 applies to the installation of bulk oxygen systems on industrial and institutional consumer premises. For the application of this section, bulk oxygen systems are defined as having a capacity greater than $13\,000\ ft^3$ of oxygen (NTP) connected in service or more than $25\,999\ ft^3$, including unconnected reserves at that site.

The major section of Subpart H, as regards the storage of laboratory chemicals, deals with flammable and combustible liquids (1910.106). This paragraph applies to the handling, storage, and use of flammable and combustible liquids with flashpoints below 200 °F. Paragraph 1910.106 includes extensive definitions [1910.106(a)]; requirements for tank storage [1910.106(b)]; requirements for piping, valves, and fittings [1910.106(c)]; and container and portable tank storage [1910.106(d)] as well as requirements for industrial plants, bulk plants, service stations, processing plants, refineries, chemical plants, distilleries, and so forth.

Subparagraph 1910.106(d) contains the most applicable requirements for storing laboratory chemicals. The scope of this paragraph covers the storage of flammable and combustible liquids in drums or other containers not exceeding 60 gallons individual capacity and portable tanks not exceeding 660 gallons individual capacity. The following points apply to containers of 60 gallons or less individual capacity. Subparagraph 1910.106(d)(2) sets forth the design, construction, and capacity of portable containers, limiting the maximum size of those containers depending on the category of flammable liquids (Classes IA, IB, IC) and combustible liquids (Classes II and III). Table H-12 of this section (found in Chapter 8 of this book) lists container size limits.

Subparagraph 1910.106(d)(3) specifies the design, construction, and capacity of storage cabinets for flammable and combustible liquids. The maximum capacity of storage cabinets for flammable liquids is 60 gallons for Class I and Class II liquids, and not more than 120 gallons for Class III liquids.

Subparagraph 1910.106(d)(4) prescribes the requirements for inside storage rooms. This passage sets requirements for storeroom construction, rating and capacity, wiring, ventilation, and storage requirements. Table H-13 in this subparagraph specifies the maximum allowable room size and quantities (gallons/ft^2) for inside storage rooms. Should drums or other containers of flammable or combustible liquids be stored elsewhere inside a building (e.g., in the receiving dock or warehouse), such storage would be subject to the requirements of Subparagraph 1910.106(d)(5); flammable or combustible liquids stored indefinitely in designated warehouses or buildings are subject to the terms of Subparagraph 1910.106(d)(5)(iv); and those stored outside buildings are subject to the conditions stated in Subparagraph 1910.106(d)(6).

Subparagraph 1910.106(d)(7) lists the requirements for fire control provisions for flammable or combustible liquids. These include the presence of portable fire extinguishers, prohibitions against open flames and smoking in such storage areas, and the prohibition of storing water-reactive materials in the same room with flammable or combustible liquids.

2.3.1.4 Fire Protection Subpart L, Paragraphs 1910.155–1910.165, specifies the OSHA requirements for fire protection for facilities. The scope of this subpart covers definitions, fire brigades, portable fire extinguishers, standpipe and hose systems, automatic sprinkler systems, fixed extinguishing systems (dry chemical, gaseous agent, water spray and foam), fire detection systems, and employee alarm systems. The regulations are followed by 19 pages of appendixes and references for fire protection.

Paragraph 1910.157 specifies requirements for the replacement, use, maintenance, and testing of portable fire extinguishers provided for the use of employees. The paragraph requires facilities (employers) to select, distribute, and provide approved portable fire extinguishers based on classes of anticipated workplace fires. Extinguishers must be made readily accessible to employees through mounting, proper location, and identification. This paragraph makes the facility (employer) responsible for inspecting, maintaining, and testing the extinguishers.

Paragraphs 1910.159 and 1910.160–1910.163 specify the requirements for fixed fire suppression equipment, such as automatic sprinkler systems and fixed extinguishing systems. Sprinkler systems installed to suppress fires of flammable and combustible liquids [1910.106(d)(7)(ii)] must comply with Paragraph 1910.159. If other fixed extinguishing systems are installed to comply with other OSHA requirements, or if such systems are installed but pose danger to the health and safety of employees, such systems are regulated by the appropriate subparagraphs of Paragraphs 1910.160–163. Paragraph 1910.161 applies to all fixed extinguishing systems using dry chemical extinguishing agents installed to meet a particular standard. Paragraph 1910.162 applies to gaseous agent fixed extinguishing agents; Paragraph 1910.163 applies to water spray and foam. Finally, all fire detection systems or employee alarm systems installed to meet the requirements of a particular OSHA standard must comply with Paragraphs 1910.164 and 1910.165.

2.3.1.5 *Electrical Requirements for Hazardous Locations*   Subpart S, Electrical, addresses electrical safety requirements that are necessary for the practical safeguarding of employees in their workplaces. This subpart prescribes the requirements for a wide variety of electrical installations.

Of particular application to laboratory chemical storage areas is Paragraph 1910.307, Hazardous (Classified) Locations. This section covers the requirements for electrical equipment and wiring for locations where flammable vapors, liquids, or gases are present. Depending on the properties of those substances, the rooms, sections, or areas are classified from most hazardous to least hazardous. Class I, Division 1 locations are the most hazardous; Class III, Division 2 locations are the least hazardous. The section defines the types of parameters for acceptable electrical installations, approved and "safe for" equipment and conduits.

2.3.1.6 *Ventilation*   OSHA specifies minimum ventilation requirements for inside storerooms of flammable or combustible liquids in Subparagraph 1910.106(d)(4)(iv). Either a gravity or a mechanical exhaust system may be used. The ventilation system must have the capability to provide a complete change of air at least six times per hour. Other provisions for switches, fresh air intake, lighting, and so on are prescribed in this section.

2.3.1.7 *Provisions for First Aid*   Subpart K, Medical and First Aid, found in Paragraph 1910.151, requires (a) the presence of readily available, physician-approved first aid supplies and (b) "suitable facilities" for quick drenching or flushing of the eyes and body within the work area for persons exposed to injurious corrosive materials.

2.3.2   Regulations for Chemical Storage Procedures

OSHA requirements for a safe workplace extend to the operations of chemical storage areas, especially their operations and maintenance, in emergency situations.

2.3.2.1   *Operations and Maintenance*   Paragraph 1910.38 specifies the requirements for Employee Emergency Plans and Fire Prevention Plans required by a particular OSHA standard. Fire prevention plans, when required, are to be written and maintained by the employer (except for employers with fewer than 10 employees) and contain elements such as a list of major workplace fire hazards and their proper handling and storage procedures, potential ignition sources and their control procedures, the type of fire protection equipment or systems to control a fire with them, job titles or names of personnel responsible for maintaining installed equipment or systems to prevent or control ignitions of fires, and the job titles or names of personnel responsible for controlling fuel source hazards. In addition, procedures must be in place to control the accumulation of flammable or combustible waste, with written housekeeping procedures. Last, employees must be apprised of the fire hazards of materials with which they work.

Subparagraph 1910.141(a)(3), Housekeeping, requires that all places of employment (e.g., chemical storage areas) be kept clean to the extent that the nature of the work allows and that the floor of the workroom be maintained in a dry condition.

Paragraph 1910.176, Handling Materials—General, specifies requirements for safe clearances when using mechanical material handling equipment; clear aisles and passageways; keeping storage areas free from accumulation of materials that constitute hazards from tripping, fire, explosion, or pest harborage; and storing materials so that their storage will not create a hazard.

2.3.2.2   *Storage of Hazardous Laboratory Wastes*   Paragraph 1910.120 outlines the conditions and requirements for worker safety and health procedures for storerooms and organizations that generate hazardous waste from laboratory operations. There are three categories of storage operations, which have different requirements under this regulation: (a) large-quantity generators who store hazardous wastes for 90 or more days; (b) large-quantity generators who store hazardous wastes for less than 90 days and small-quantity generators who have emergency response teams (as defined in Section 2.3.2.5 of this chapter); and (c) large-quantity generators who store hazardous wastes for less than 90 days and small-quantity generators who do not have emergency response teams.

Category c, the specified generators who do not have emergency response teams, are exempt from the requirements of Paragraph 1910.120.

Category b organizations must comply with Subparagraph (p)(8) of Paragraph 1910.120. This section requires an emergency response program that includes an emergency response plan (with specified elements), training, and procedures for handling emergency incidents.

Category a organizations, large-quantity generators who store hazardous waste for more than 90 days, must comply with the entire requirements of Subparagraph (p) of Paragraph 1910.120. These requirements include (a) a safety and health program; (b) a hazard communication program; (c) a medical surveillance program; (d) a decontamination program; (e) a new technology program (for laboratory hazardous waste storage operations limited to new industry control measures in spill control); (f) a material handling program [limited to requirements of Subparagraphs (j)(1)(ii) through (vii) and (xi), (j)(3), and (j)(8)]; (g) a training program, with specific provisions for new and current employees as trainers providing the training; and (h) an emergency response program. (EPA requirements for the storage of hazardous waste are highlighted in Section 2.4.1 of this chapter.)

2.3.2.3 *Emergency Response*   The final rule of Hazardous Waste Operations and Emergency Response, Paragraph 1910.120, published on March 6, 1989, outlines the requirements for safety and health in emergency response operations for releases, or substantial threats of releases, of hazardous substances regardless of the location of the hazard. These requirements affect chemical storage operations when employees (the HAZMAT team) come from outside the immediate release area to deal with what is or may be an uncontrollable release of a hazardous chemical. If employees in the immediate area can control an incidental release of a hazardous chemical, these efforts are not considered to be emergency responses according to the standard and therefore are not regulated. If other employees, designated to respond to emergency releases, are called into that area to deal with the chemical release, then the provisions of the standard must be met.

Subparagraph (q) of Paragraph 1910.120 describes the requirements for emergency response to hazardous substances:

1. Development and implementation of an Emergency Response Plan [not required for organizations that do not use their own employees for emergency response, provided they have an emergency action plan (1910.38(a)]

2. Specific elements of an Emergency Response Plan, including pre-emergency planning; personnel roles, lines of authority, training, and communication; emergency recognition and prevention; safe distances and places of refuge; site security and control; evacuation routes and procedures; decontamination; emergency medical treatment with first aid; emergency alerting and response procedures; critique of response and follow-up; personal protective equipment and emergency response equipment

3. Procedures for handling emergency response, including the designation and operation of an incident command system

4. Cooperation and involvement of skilled support personnel

5. Involvement of specialist employees for technical support
6. Training of the response personnel
7. Specifications for trainers
8. Specifications for refresher training of response personnel
9. Medical surveillance and consultation for the response team
10. Standards for chemical protective clothing
11. Postemergency response operations, including cleanup

Appendix B of Paragraph 1910.120 discusses the levels of protection and protective gear to help comply with number 10 listed above. (Further information on OSHA requirements for emergency response is given in Chapter 6.)

2.3.3 Regulations for Protecting People

2.3.3.1 Personal Protective Equipment A proposed revision (August 1989) to Subpart I, Personal Protective Equipment, Paragraphs 1910.132–1910.133, 1910.135, and 1910.136, requires employers to select the types of personal protective equipment that will protect employees from the particular occupational hazard(s) likely to be encountered. Employers are responsible for training employees in the proper use of personal protective equipment. The most applicable section to chemical storage operations is Paragraph 1910.133, which deals with eye and face protection, specifically from liquid chemicals, acid/caustic liquids, and chemical gases or vapors. Eye and face protection must properly fit the employee. Paragraph 1910.132 references ANSI Z87.1-1989, "Practice for Occupational and Educational Eye and Face Protection."

The proposed revision to Subpart I prescribes the proper selection of personal protective equipment based on a hazard assessment of the workplace. Appendix B of the proposed rule (*Federal Register* **54**(157), 33843–33844) contains nonmandatory guidelines for assessing personal protective equipment selection. For operations where chemicals (acid and chemical handling) are the source, protection against splashes by the use of goggles, eyecups, or faceshields is recommended. Faceshields should only be worn over primary eye protection (e.g., goggles). Eye and face protection should be chosen that provides adequate ventilation but still protects the wearer from splash entry.

Paragraph 1910.134 specifies the requirements for the use of respirators as personal protective equipment. The paragraph requires the use of respirators to protect workers from air "contaminated with harmful dusts, fogs, mists, fumes, smokes, gases, sprays or vapors" where effective engineering controls are not feasible. The paragraph requires a written respirator program with standard operating procedures governing the selection and use of respirators, training of users, and regular maintenance, cleaning, and disinfection.

Workers must be examined to determine their physical capability to wear a respirator before it can be assigned. The paragraph specifies requirements for the fitting and use of airline respirators, self-contained breathing apparatus, gas masks, and other respirators.

2.3.3.2 *First Aid and Medical Services*   Paragraph 1910.151 states the requirements for first aid: the presence of a designated person to administer first aid, and availability of medical consultation and emergency water to aid those exposed to corrosive materials.

2.3.3.3 *Emergency Action Plans*   Subparagaph 1910.38(a) applies to all emergency action plans required by a particular OSHA standard. Emergency action plans are to be written, covering the designated actions both employers and employees must take to ensure employee safety during fires and other emergencies. The elements of an emergency action plan include emergency escape procedures, procedures for employees remaining to operate critical functions before evacuation, procedures to account for employees after evacuation, rescue and medical duties, preferred means of reporting fires and emergencies, and personnel to contact for further information. The section also requires the establishment of an employee alarm system and the training of employees in emergency action procedures.

2.3.4 Regulations Involving Information

2.3.4.1 *Hazard Communication*   On August 24, 1987, OSHA published the revised Hazard Communication Standard (Paragraph 1910.1200), which requires all employers to establish hazard communication programs to transmit information on the hazards of chemicals to their employees by specified means. The burden for meeting the requirements of this law lies with the employer. There are two distinctions that apply to chemical storage operations: (a) chemical storage operations that inventory and supply sealed containers of hazardous chemicals, and (b) chemical storage operations, like dispensing, in which an employee has the potential for exposure to the chemical under normal working conditions.

Pertaining to supply of sealed containers, employers must (a) ensure that labels on incoming containers of hazardous chemicals are not defaced or removed; (b) maintain and make available Material Safety Data Sheets to employees as requested; and (c) provide information and training on how employees can protect themselves in the event of spills or leaks from sealed containers.

In addition, for storeroom operations in which hazardous chemicals are normally handled outside of sealed containers (as in dispensing), employers must (d) provide a written hazard communication program when chemical storage is an operation independent of laboratory operations and (e) ensure that each container of hazardous chemicals is labeled with the identity and appropriate hazard warnings of the chemicals when it passes from the storage area to the laboratory for use.

The Hazard Communication Standard specifies the information to be included on Material Safety Data Sheets and labels. In addition, employers must provide employees with information on (f) the requirements of the Hazard Communication Standard, (g) any operations in the work area in which hazardous chemicals are present, and (h) the location and availability of the written hazard communication program.

Employers must provide training to employees prior to their work with hazardous chemicals and on an annual basis. At a minimum, this training is to include (a) methods and observations that may be used to detect the presence or release of a hazardous chemical in the work area, (b) physical and health hazards of chemicals in the work area, (c) protective measures from chemical hazards, and (d) details of the hazard communication program developed by the employer, including the labeling system and Material Safety Data Sheets, and how to obtain and use hazard information.

2.3.4.2 Occupational Exposure to Hazardous Chemicals in Laboratories

The final rule of Paragraph 1910.1450, the Laboratory Standard, was published on January 31, 1990, with an effective date of January 31, 1991. The rule applies to all employers engaged in the laboratory use of chemicals with the intent to limit occupational exposure to hazardous chemicals in laboratory operations. Although the major part of this rule applies to those exposures, the rule has implications for the chemical storage that occurs in laboratories. Subparagraph (f), Employee Information and Training, requires that "employers provide employees with information and training to ensure that they are apprised of the hazards of chemicals present in their work area." Subparagraph (f)(3), Information, states that employees be informed of "the location and availability of known reference material on the hazards, safe handling, storage and disposal of hazardous chemicals found in the laboratory." Subparagraph (f)(4)(b), Training, stipulates that employee training shall include "the physical and health hazards of chemicals in the work area." Physical hazards are defined as those "chemical[s] for which there is a scientifically valid evidence that [it] is a combustible liquid, a compressed gas, explosive, flammable, an organic peroxide, an oxidizer, pyrophoric, unstable (reactive) or water-reactive."

Chemicals are stored in laboratories. Improper storage practices and conditions can create additional physical hazards that initiate the release of hazardous energy (see Chapter 5). The implication of this section, according to this author's interpretation, is that laboratory workers should know the storage conditions that are related to the physical hazards of these chemicals. Improper chemical storage conditions and procedures have and can lead to fires, explosions, and releases of chemicals that can have adverse safety and health effects. This means that laboratory workers involved with the storage of chemicals in the laboratory should know and establish the safe storage conditions and environment for laboratory storage as well as follow safe chemical storage procedures.

Another pertinent subparagraph to this standard is Subparagraph 1910.1450(h), Hazard Identification, which states that "employers shall ensure that labels on incoming containers of hazardous chemicals are not removed or defaced" and "employers shall maintain any material safety data sheets that are received with incoming shipments of hazardous chemicals, and ensure that they are readily accessible to laboratory employees."

Appendix A to the standard contains National Research Council recommendations concerning chemical hygiene in laboratories, which is non-mandatory but included as a guide for developing a chemical hygiene plan. Those recommendations include sections on stockrooms/storerooms and laboratory storage.

2.3.4.3 *Physical Hazard Marking and Accident Prevention Signs/Tags*   Paragraph 1910.144, Safety Color Code for Marking Physical Hazards, specifies the use of the color red for identification of fire protection equipment apparatus, danger, and stop. Yellow is the basic color for designating caution and for marking physical hazards, such as striking against, stumbling, falling, tripping, and so on.

Paragraph 1910.145, Specifications for Accident Prevention Signs and Tags, states the requirements for the design, application, and use of signs or symbols to indicate or define hazards that may lead to accidental injury of workers, the public, or both, or property damage. Signs are classified as Danger, Caution, and Safety Instruction. Employers are to instruct employees on the significance of the sign and what it indicates. The section also sets forth the requirements for using accident prevention tags, which are used in extraordinary circumstances where signs are not readily available or posted. Where chemical storage areas or operations have the potential for Danger, Caution, or Safety Instruction, signs (and possibly tags) are to be posted in compliance with this section.

2.3.4.4 *Permissible Exposure Limits for Air Contaminants*   Subpart Z, Paragraphs 1910.1000–1910.1500, regulates worker exposure to toxic and hazardous substances. Paragraph 1910.1000, also known as the Z tables, specifies the permissible exposure limit (PEL), or the maximum legal concentration to which a worker may be exposed for an 8-hour period. The PELs are expressed in PPM (parts per million) for vapors and gases, and mg/m^3 (milligrams/cubic meter) for particulates. The initial Z tables were promulgated in the early 1970s and were based on 1968 ACGIH threshold limit values, covering a total of 428 substances. The revised standard was published in March 1989, with a compliance date of September 1989, and includes updated PELs for 212 of the original 428 substances. PELs for 164 new substances were added to the tables. Table Z-1 contains transitional limits and final rule columns, which specify the interim and final maximum allowable concentrations of

these substances. Table Z-2 contains the acceptable ceiling concentration and maximum peak above the ceiling for 21 chemical compounds (including benzene, carbon disulfide, methylene chloride, and toluene). The final compliance date for using engineering controls to keep concentrations below the PEL has been set as December 31, 1992.

While the routine storage of laboratory chemical containers does not expose a worker to dangerous concentrations, storeroom operations such as dispensing or repackaging can cause vapor formation that can expose a worker to inhalation or skin contact hazards.

Subpart Z also contains standards for 20 individual chemicals in separate paragraphs, including:

Cancer Suspect Agents, such as 4-nitrobiphenyl (1910.1003), *a*-naphthylamine (1910.1004), methyl chloromethyl ether (1910.1004), 3,3'-dichlorobenzidene (1910.1007), bis(chloromethyl) ether (1910.1008), *b*-naphthylamine (1910.1009), benzidine (1910.1010); 4-aminodiphenyl (1910.1011), ethyleneimine (1910.1012), *b*-propiolactone (1910.1013), 2-acetylaminofluorine (1910.1013), 4-dimethylaminoazobenzene (1910.1014), *N*-nitrosodimethyl amine (1910.1014), and vinyl chloride (1910.1017)

Cancer Hazards, such as inorganic arsenic (1910.1018), benzene (1910.1028), 1,2-dibromo-3-chloropropane (1910.1044), and acrylonitrile (1910.1045)

Cancer and Reproductive Hazard, ethylene oxide (1910.1047)

Irritant/Potential Cancer Hazard, formaldehyde (1910.1048)

These chemicals must be appropriately labeled to reflect these hazards as specified in the individual standards. Furthermore, storage suggestions are provided in Appendix A (Substance Safety Data Sheet) of that paragraph for most of these chemicals. (These suggestions may be mandatory or nonmandatory depending on the standard.)

2.4 EPA REGULATIONS

The Environmental Protection Agency (EPA) was formed by order of the President under the passage of the Environmental Protection Act, signed into law in the 1970s. The EPA has the responsibility to develop and promulgate regulations to protect the environment from contamination by hazardous chemicals or other hazards. This mission is reflected in two basic approaches: (a) response to disasters done to the environment in the past (e.g., Love Canal) and (b) prevention of future damage to the environment by regulating current and future activities.

The Comprehensive Environmental Response, Compensation and Liability Act (CERCLA, also known as Superfund) was passed in 1980 to respond to past damage by establishing a National Priority List and cleanup schedule of uncontrolled hazardous waste sites, among other actions. In 1986, Superfund was reauthorized as the Superfund Amendment and Reauthorization Act (SARA), which included broader provisions for reporting inventories and releases of hazardous substances.

Other acts, such as the Clean Air Act, Resource Recovery and Conservation Act (RCRA), and the Emergency Planning and Community Right-to-Know Act, have established regulations to prevent future damage to the environment by regulating present and future activities with such potential.

The number system for EPA regulations is structured similarly to OSHA regulations. The entire set of EPA regulations is codified in 40 CFR and organized into chapters, subchapters, and parts. EPA regulations are updated as needed in the daily *Federal Register.*

The EPA regulations discussed here are drawn from those dealing with Hazardous Waste (40 CFR Parts 160–265), Emergency Planning and Notification (40 CFR Part 355), and Hazardous Substance Reporting (40 CFR Part 370). These regulations refer to the appropriate title, subtitle, and sections of the published laws (RCRA, SARA, or other public laws; e.g., RCRA Title II, Subtitle C, Section 3001 deals with Listing of Solid Wastes).

EPA regulations affect the storage of laboratory chemicals in several ways. The following sections discuss the requirement from the preventive posture of RCRA (geared toward the storage of hazardous wastes) and appropriate requirements of SARA, especially the Emergency Planning and Community Right-to-Know regulations.

2.4.1 Storing Hazardous Waste

Although there are specific regulations for the management of hazardous wastes that involve storage, transportation, treatment, and disposal, this section will discuss only storage requirements. EPA requirements for managing hazardous waste can be difficult to determine. To understand how the regulations apply to your operations, you must examine the criteria for compliance published in the standards and then see whether your operations or organization falls under the scope and application of the law. Requirements for the storage of hazardous waste generated from laboratory chemicals depend on your generator status. There are three groups of hazardous waste generators, as shown in Table 2.1. Notice that the driving force for determining generator status is quantity generated within a defined time period. Acutely hazardous wastes and hazardous wastes are listed in Part 261, either by specific name or identifying characteristics (such as ignitability, corrosivity, reactivity, and toxicity). The generator status is determined by counting the quantity of hazardous waste streams for a calendar month.

2.4.1.1 *Conditionally Exempt Small-Quantity Generators* Section 261.5 sets forth the requirements of conditionally exempt small-quantity generators. Conditionally exempt small-quantity generators may accumulate acutely

TABLE 2.1 Decision Logic for Determining EPA Generator Status of Hazardous Waste

Source	Amount of Hazardous Waste Generated in One Month	
	Acutely Hazardous (kg)	Hazardous (kg)
Small-quantity generator (conditionally exempt)	<1	<100
Small-quantity generator	<1	<1000
Generator	>1	>1000

hazardous wastes and nonacute hazardous wastes on-site with no time limit. However, if all acutely hazardous waste totals more than 1 kg in any month, or if more than 1000 kg of hazardous waste is accumulated on-site at any one time, all EPA regulations for hazardous waste must be met. Conditionally exempt small-quantity generators must comply with Section 262.11, which requires the determination of hazardous waste through listing, identifying characteristics, or testing. Section 262.11 also requires the conditionally exempt small-quantity generator to refer to Part 264, 265, or 268 for possible exclusions or restrictions to the regulations for the management of hazardous waste.

2.4.1.2 Small-Quantity Generators Many laboratories and laboratory chemical storage operations qualify for the small-quantity generator status. Small-quantity generators have to meet more storage requirements than those that are conditionally exempt, but far fewer than those with generator status. Small-quantity generators may accumulate hazardous waste on-site for up to 180 days without having a permit or interim status if (a) total on-site accumulation never exceeds 6000 kg; (b) containers of hazardous waste are used and managed in compliance with Sections 265.170–175, and 177 (see Table 2.2); (c) each container of hazardous waste is clearly marked with the words HAZARDOUS WASTE, along with the date of initial accumulation of wastes; (d) the requirements of Section 265.201 are met if hazardous waste is stored in tanks; and (e) there is an emergency coordinator and the procedures listed in Subsection 262.34(d) are followed. A small-quantity generator may accumulate hazardous waste for up to 270 days on-site if the waste is to be transported to a disposal site more than 200 miles away, as long as the above conditions are met. If a small-quantity generator accumulates more than 6000 kg of hazardous waste or exceeds the 180/270-day accumulation period, that generator automatically becomes an operator of a storage facility, is subject to 40 CFR Parts 264 and 265, and must follow the permit requirements of Part 270.

TABLE 2.2 Use and Management of Containers for Storing Hazardous Waste

Requirement	Small-Quantity Generator	Generator
Each container marked with HAZARDOUS WASTE [262.34]	X	X
Each container marked with initial date of accumulation [262.34]	X	X
Hazardous waste is stored in nonleaking containers in good condition [265.171]	X	X
Hazardous waste in leaking or deteriorating containers is transferred to containers in good condition [265.171]	X	X
Containers for holding hazardous waste are made of or lined with materials compatible with the waste [265.172]	X	X
Containers are always closed, except when adding or removing waste [265 173(a)]	X	X
Containers holding waste must not be stored in a manner to rupture the container or cause it to leak [265.173(b)]	X	X
Storage areas for containers of hazardous waste are inspected at least weekly [265.174]	X	X
Containers holding ignitable or reactive waste must be located at least 15 meters (50 ft) from the property line [265.176]		X
Incompatible wastes must not be placed in the same container [265.177(a)]	X	X
Hazardous waste must not be placed in an unwashed container previously holding an incompatible waste or material [265.177(b)]	X	X
Storage containers holding wastes incompatible with any wastes or materials stored nearby are separated or protected by dike, berm, wall, or other device [265.177(c)]	X	X

2.4.1.3 Generators Generators may accumulate hazardous waste on-site for up to 90 days without having a permit or interim status if (a) waste is placed in containers or tanks that are used and managed in accordance with Sections 265.170–177 (see Table 2.2) or Subpart J (for tanks) of Part 265, (b) containers or tanks are marked with both the words HAZARDOUS WASTE and the date of initial accumulation, and (c) the generator complies with the requirements in Subparts C and D in 40 CFR Part 265 and Section 265.16. Subpart C of Part 265 specifies safety and preparedness equipment, such as internal and external communications, portable fire extinguishers, and spill control and decontamination equipment, and availability of water hoses. Subpart D requires written contingency, spill control countermeasures, and emergency plans. If a generator accumulates hazardous waste for more than 90 days, that generator becomes an operator of a storage facility subject to all the requirements of Parts 264, 265, and the permit requirements of Part 270. An extension of up to 30 days may be granted by the regional administrator for unforeseen, temporary, or uncontrollable circumstances.

2.4.1.4 *Owner/Operators of Storage Facilities* Any generator or small-quantity generator who exceeds either the on-site accumulation limit of 6000 kg of hazardous waste or the respective on-site accumulation time is automatically designated as an owner/operator of a storage facility. This status entails much more stringent requirements, specified either by Part 264 or 265 (interim status) and by Part 270 (permits), which include facility design, equipment, training, emergency plans, reporting of hazardous wastes stored, and so on. A description of these specific requirements is outside the scope of this chapter; concerned readers should refer to those parts.

2.4.2 Emergency Planning and Notification

The Superfund Amendment and Reauthorization Act of 1986 (SARA) authorized EPA to issue regulations for organizations storing specific amounts of hazardous substances. The act was influenced by the Bhopal incident, in which hundreds of citizens were killed or injured by a release of toxic gas. Part 355 of 40 CFR outlines the requirements for emergency planning and notification for organizations with hazardous substances. This part specifies that a facility, if it has amounts equal to or more than the threshold planning quantity (TPQ) of any substance listed in its Appendix A or B (or designated by the state) on-site at any one time, must (a) notify the local emergency planning committee; (b) provide a facility emergency coordinator; (d) provide information regarding any changes in quantities or local emergency plans; (e) notify the community emergency coordinator (or local emergency response personnel) or state emergency response commission if a release will affect either the local or state community; and (f) provide the specified information about the released substance in (e).

There are two notable provisions for the storage of laboratory chemicals. First, the regulations for reporting releases do not apply if the release is confined within a facility's property boundaries and the only exposure is to the employees at the facility. Second, notification to the local or state emergency planning committees depends on inventory surpassing the TPQ at any one time. TPQs are relatively high amounts for normal laboratory chemical storage operations. For example, the TPQ for carbon disulfide is 10000 lb, for chlorine, 100 lb, for sulfuric acid, 1000 lb. Readers should examine their inventory of specific laboratory chemicals and compare them with the TPQs listed in Appendixes A and B of Part 355 to determine the application of this section. Any substance used in a research laboratory, hospital, or medical laboratory under the direct supervision of a technically qualified individual is excluded from compliance with this section.

2.4.3   Hazardous Substance Reporting

2.4.3.1   General   Another requirement of SARA is that facilities must provide the public with information on hazardous chemicals to enhance community awareness of chemical hazards and facilitate the development of state and local emergency response plans. In 40 CFR 370, the regulations call for submitting Material Safety Data Sheets (MSDSs), and inventory (Tier 1 and/ or Tier 2) reports for all hazardous substances or hazardous chemicals present at the facility that exceed minimum threshold levels. The minimum threshold levels are specified in Section 370.20. The MSDSs and inventory reports must be submitted to (a) the state emergency response commission, (b) the local emergency planning committee (LEPC), and (c) the fire department having jurisdiction over the facility (hereafter referred to as the local fire department).

2.4.3.2   MSDS Reporting   As of October 1989 MSDSs must be submitted for all hazardous chemicals (as defined by OSHA's Hazard Communication list in 29 CFR 1910.1200) or extremely hazardous substances (listed in 40 CFR 355), regardless of quantity. As an alternative to submitting an MSDS for each hazardous chemical, a facility may submit a categorized list of chemicals with the chemical or common name listed on the MSDS. There are five categories for this list:

1. *Immediate (acute) health hazard:* highly toxic, toxic, irritant, sensitizer, corrosive
2. *Delayed (chronic) health hazard:* includes carcinogens
3. *Fire hazard:* flammable liquids, combustible liquids, pyrophorics, oxidizers
4. *Sudden release of pressure:* explosives, compressed gases
5. *Reactive:* unstable reactive, organic peroxide, water reactive

Revised MSDSs must be provided to the appropriate committee within three months after they are received. If only a list of chemicals is provided to the committees, an MSDS must be provided within 30 days of the receipt of a request.

2.4.3.3   Inventory Reporting   Section 370.25 specifies the requirements for inventory reporting. Each facility that qualifies must provide an inventory form to the state, LEPC, and local fire department on the hazardous chemicals present at the facility during the preceding year. The inventory form must be submitted annually, by March 1, using the Tier 1 format (Figure 2.1). After March 1990, or the third year in which the facility becomes subject to this regulation, Tier 1 reports are required for all hazardous chemicals in any

Page ____ of ____ pages
Form Approved OMB No. 2050-0072

Tier One **EMERGENCY AND HAZARDOUS CHEMICAL INVENTORY**
Aggregate Information by Hazard Type

FOR OFFICIAL USE ONLY — ID # ____ Date Received ____

Important: Read instructions before completing form

Reporting Period From January 1 to December 31, 19____

Facility Identification

Name ____
Street Address ____
City ____ State ____ Zip ____
SIC Code ☐☐☐☐ Dun & Brad Number ☐☐-☐☐☐-☐☐☐☐

Owner/Operator

Name ____
Mail Address ____
Phone ( ) ____

Emergency Contacts

Name ____
Title ____
Phone ( ) ____
24 Hour Phone ( ) ____

Name ____
Title ____
Phone ( ) ____
24 Hour Phone ( ) ____

☐ Check if site plan is attached

	Hazard Type	*Max Amount**	*Average Daily Amount**	*Number of Days On-Site*	*General Location*
Physical Hazards	Fire	☐☐	☐☐	☐☐☐	____
	Sudden Release of Pressure	☐☐	☐☐	☐☐☐	____
	Reactivity	☐☐	☐☐	☐☐☐	____
Health Hazards	Immediate (acute)	☐☐	☐☐	☐☐☐	____
	Delayed (Chronic)	☐☐	☐☐	☐☐☐	____

Certification *(Read and sign after completing all sections)*

I certify under penalty of law that I have personally examined and am familiar with the information submitted in this and all attached documents, and that based on my inquiry of those individuals responsible for obtaining the information, I believe that the submitted information is true, accurate and complete

Name and official title of owner/operator OR owner/operator's authorized representative ____

Signature ____ Date signed ____

* Reporting Ranges

Range Value	Weight Range in Pounds From...	To...
00	0	99
01	100	999
02	1000	9,999
03	10,000	99,999
04	100,000	999,999
05	1,000,000	9,999,999
06	10,000,000	49,999,999
07	50,000,000	99,999,999
08	100,000,000	499,999,999
09	500,000,000	999,999,999
10	1 billion	higher than 1 billion

FIGURE 2.1 Tier 1 inventory report for hazardous chemicals.

quantity and for any extremely hazardous substance with specific quantities. The quantity levels for extremely hazardous substances which trigger reporting are (a) 500 or more pounds, (b) 55 gallons, or (c) the TPQ, whichever is less.

A facility may elect to submit the Tier 2 (Figure 2.2) form instead of the Tier 1 report. Tier 2 provides much more detailed information on the actual location, type of containers, and inventory parameters for each hazardous chemical or extremely hazardous substance. After Tier 1 reports or a categorized list are submitted, the state, local committee, or fire department may request Tier 2 information at any time. The facility must provide the report within 30 days of the request. The facility must also allow an on-site inspection by the local fire department upon request of that department. Sections 40 CFR 370.40 and 370.41 display the Tier 1 and Tier 2 forms and instructions for completion.

2.4.3.4 Public Access to Information   Any person may obtain an MSDS with respect to a specific facility by submitting a written request to the local emergency planning committee. Tier 2 information requested in writing by any person must be made available to the requester. The specific location of the on-site chemicals can be withheld from the requester by the owner/operator of the facility. If unavailable, the state or local committee must request Tier 2 information from the facility if the written request is from a state or local official acting in an official capacity, or if the written request is limited to hazardous chemicals stored at the facility in amounts exceeding 10000 pounds. If the request is from the general public on chemicals stored in less than 10000 pounds, the state may request a Tier 2 form if the written request includes a general statement of need.

2.5 ROLE OF STATE AND LOCAL REGULATIONS

State and local regulations may have an additional impact on the storage of laboratory chemicals. This is especially true when state and local governments adopt different or newer versions of national consensus standards that are more stringent than the ones referenced by federal regulations. For example, effective July 1991, the state of New York will require all schools to adopt safety and security measures for chemical storage, based on the guidelines promulgated by the state's department of education. Local building and occupancy codes may require specific construction, design, and/or operating protocols where hazardous chemicals are stored. Each state may have additional statutes or regulations promulgated through its departments of labor or environmental protection agency. Some state plans adopt only the minimum federal regulations; others may enhance the federal regulations with stricter compliance requirements. A prudent operations manager will consult with state and local authorities to determine the extent of their legal requirements for storing laboratory chemicals.

2.6 CONCLUSION

OSHA and EPA regulations have an impact on laboratory chemical storage operations. These federal regulations promulgate specific requirements, which involve state and local enforcement for compliance.

This chapter has illustrated a number of applicable standards that affect chemical storage operations. It has discussed OSHA's General Industry Standards and applications to storage operations for laboratory chemicals, as well as EPA hazardous waste and notification regulations for hazardous chemicals and extremely hazardous substances.

Federal regulations are dynamic. The *Federal Register* regularly lists new regulations and changes in current regulations, and announces proposed requirements. A proactive stance can help an organization anticipate and influence the impact of federal regulations. The following suggestions indicate how an organization can be proactive in this area:

1. *Be informed.* Investigate your organization's legal liability under the scope and application of these regulations. Remediate those areas where compliance is lacking.
2. *Keep abreast of new regulatory activity.* Know where the regulatory agencies are going with proposed rules. Read the *Federal Register* or qualified newsletters on OSHA and EPA activity regularly.
3. *Be an early adopter of new standards.* New and revised national consensus standards form the basis for tomorrow's laws. Join a standards subcommittee and help determine the future of compliance.
4. *Get involved with the rule-making process.* OSHA and EPA are somewhat flexible in accepting input from industry and the regulated community. Attend hearings, submit written comments, offer testimony where appropriate, and/or join a lobbying group or association that represents the common interests of chemical safety (such as the American Chemical Society).
5. *Investigate the possibility of variances.* In special instances OSHA and EPA may grant a variance or exemption from specific requirements if it is demonstrated that an alternative is as safe and economical as what is required by the regulation.

Federal, state, and local regulations offer the opportunity to achieve a safer workplace and environment. How that opportunity is grasped is up to you.

REFERENCES

Environmental Statutes, 10th ed., Government Institutes, Inc., Rockville, MD, 1989.
General Industry Standards Code of Federal Regulations, 29 CFR 1910, US Department of Labor, Government Printing Office, Washington, DC, July 1989.

Tier Two

EMERGENCY AND HAZARDOUS CHEMICAL INVENTORY

Specific Information by Chemical

Facility Identification

Name _____
Street Address _____
City _____ State _____ Zip _____

SIC Code [ ] Dun & Brad Number [ ]

FOR OFFICIAL USE ONLY ID # _____ Date Received _____

Owner/Operator Name

Name _____
Mail Address _____

Emergency Contact

Name _____ Title _____
Phone ( ) _____ 24 Hr. Phone ( ) _____

Name _____ Title _____
Phone ( ) _____ 24 Hr. Phone ( ) _____

Phone ( ) _____

Important: Read all instructions before completing form

Reporting Period From January 1 to December 31, 19 ___

Chemical Description

	Physical and Health Hazards (check all that apply)	Inventory	Storage Codes and Locations (Non-Confidential)

CAS [ ] [ ] [ ] [ ] [ ] Trade Secret [ ]
Chem. Name _____

Check all that apply: Pure [ ] Mix [ ] Solid [ ] Liquid [ ] Gas [ ]

Fire [ ]
Sudden Release of Pressure [ ]
Reactivity [ ]
Immediate (acute) [ ]
Delayed (chronic) [ ]

Max. Daily Amount (code) [ ]
Avg. Daily Amount (code) [ ]
No. of Days On-site (days) [ ]

Storage Code   Storage Locations

CAS [ ] [ ] [ ] [ ] [ ] Trade Secret [ ]
Chem. Name _____

Check all that apply: Pure [ ] Mix [ ] Solid [ ] Liquid [ ] Gas [ ]

Fire [ ]
Sudden Release of Pressure [ ]
Reactivity [ ]
Immediate (acute) [ ]
Delayed (chronic) [ ]

Max. Daily Amount (code) [ ]
Avg. Daily Amount (code) [ ]
No. of Days On-site (days) [ ]

CAS [ ] [ ] [ ] [ ] [ ] Trade Secret [ ]
Chem. Name _____

Check all that apply: Pure [ ] Mix [ ] Solid [ ] Liquid [ ] Gas [ ]

Fire [ ]
Sudden Release of Pressure [ ]
Reactivity [ ]
Immediate (acute) [ ]
Delayed (chronic) [ ]

Max. Daily Amount (code) [ ]
Avg. Daily Amount (code) [ ]
No. of Days On-site (days) [ ]

Certification (Read and sign after completing all sections)

I certify under penalty of law that I have personally examined and am familiar with the information submitted in this and all attached documents, and that based on my inquiry of those individuals responsible for obtaining the information, I believe that the submitted information is true, accurate, and complete

Name and official title of owner/operator OR owner/operator's authorized representative

Signature _____   Date signed _____

Optional Attachments (Check one)

[ ] I have attached a site plan
[ ] I have attached a list of site coordinate abbreviations

32

FIGURE 2.2 Tier 2 inventory report for specific information on hazardous chemicals.

33

Hall, J., Jr., et al. (1985). *RCRA Hazardous Waste Handbook,* 6th ed Government Institutes, Inc., Rockville, MD.

Lowry, G. and R. C. Lowry (1986). *Handbook of Hazard Communication & OSHA Requirements,* Lewis, Chelsea, MI.

Occupational Exposures to Hazardous Chemicals in Laboratories, Final Rule Code of Federal Regulations, 29 CFR 1910.1450 *Federal Register,* Government Printing Office, Washington, DC, Wednesday, January 31, 1990.

Protection of the Environment, Code of Federal Regulations, 40 CFR 260-370, Environmental Protection Agency, Government Printing Office, Washington, DC, July 1989.

Subpart I—Personal Protective Equipment, Proposed Rule Code of Federal Regulations, *Federal Register,* Government Printing Office, Washington, DC, Wednesday, August 16, 1989.

CHAPTER 3

STORAGE REQUIREMENTS FOR FLAMMABLE AND HAZARDOUS CHEMICALS

NORMAN V. STEERE
Laboratory Safety and Design Consultant
Norman V. Steere & Associates, Inc.
Minneapolis, Minnesota

3.1 INTRODUCTION

Storage of chemicals should minimize safety and health hazards to personnel, equipment, buildings, and the environment. Safe storage will require appropriate construction, equipment, and operating practices. Chemical containers should be stored and ventilated to prevent breakage or leakage which might endanger someone who enters or works in the storage area or which might cause deterioration of the chemicals or damage to containers, equipment, or the building. Facilities for storage of chemicals outside of laboratory and research areas should be separated and protected so that a fire or spill in such stockrooms or storage areas is not likely to spread beyond the room.

Safe storage of hazardous chemicals is based partly on meeting the requirements of codes and standards which are based on adverse experience incorporated into codes and regulations. Codes, standards, and code-enforcing authorities can provide some important guidance, but they usually cannot address all the safety and health aspects of chemical storage.

Safe storage is also based on designing facilities and equipment that provide for the control of recognized hazards that are not yet part of codes or standards, particularly as chemicals with different or greater hazards come into more common use.

This chapter describes and discusses some of the more common requirements and needs for safe storage of chemicals within a laboratory and in a separate room within a building used for teaching, research, testing, or other laboratory activities. The chapter also discusses needs for storage cabinets, safe dispensing, and emergency procedures. Emphasis is placed on practical ways of storing hazardous chemicals, with references to storage requirements specified in the standards used by local or state fire authorities, insurance inspectors, the Occupational Safety and Health Administration and other federal agencies.

Several National Fire Protection Association (NFPA) standards will be referred to:

Flammable and Combustible Liquids Code, NFPA 30-1990

Hazardous Chemicals Data, NFPA 49-1991

Code for Storage of Liquid and Solid Oxidizing Materials, NFPA 43A-1990

Code for Storage of Gaseous Oxidizing Materials, NFPA 43C-1986

Code for Storage of Pesticides in Portable Containers, NFPA 43D-1986

Fire Protection Standard for Laboratories Using Chemicals, NFPA 45-1986

Standard for Health Care Facilities, NFPA 99-1990

Uniform Fire Code 1988 and Uniform Building Code 1988

Storage of cylinders of acetylene, fuel gases, hydrogen, and oxygen may be regulated by specific NFPA codes, depending on the sizes of containers and the total quantities stored.

While NFPA standards are widely used as regulations by political subdivisions that adopt them by reference, the Uniform Building Code and the Uniform Fire Code are used only in the western United States and other states that adopt them by reference. The 1988 editions of the Uniform Building Code and the Uniform Fire Code include a great many specific requirements for storage and dispensing of hazardous materials, including spill control, drainage, containment, ventilation, and fire protection. The Uniform Fire Code (UFC) requires a storage plan for all storage facilities, including the location and dimensions of aisles and the intended storage arrangement.

Table 3.1 lists the quantities of hazardous materials exempt from the requirements of the UFC. Greater quantities require a permit and compliance with special storage requirements. For example, the 1988 UFC requires automatic sprinklers for some storage rooms, contains segregation requirements for hazardous chemicals, and specifies storage conditions. For storage of toxic compressed gases or highly toxic gases, liquids, or solids, the UFC requires that a minimum of two self-contained breathing apparatuses be provided nearby in a safe area.

3.2 GENERAL RECOMMENDATIONS

Safe storage of chemicals must begin with identification of the chemicals to be stored and their hazards. Since many chemicals have several hazards

TABLE 3.1 Hazardous Materials Regulated by 1988 Uniform Building Code and Uniform Fire Code[a]

Hazardous Material	Class	Storage	Use	
			Closed Systems	Open Systems
Combustible liquid	II	120	120	30
	IIIA	330	330	80
Cryogenic liquid, flammable or oxidizing		45	45	10
Flammable solid		125	25	25
Flammable liquid	IA	30	30	10
	IB	60	60	15
	IC	90	90	20
Combination		120	120	30
Oxidizer	4	1	¼	¼
	3	10	2	2
	2	250	250	50
	1	1000	1000	200
Gaseous		1500	1500	—
Pyrophoric		4; 50 ft^3	1; 10 ft^3	0
Corrosives		5000 lb, 500 gal, 650 ft^3	Same	1000 lb, 100 gal
Highly toxics		1 lb, 20 ft^3	Same	¼ lb
Irritants		5000 lb, 500 gal, 650 ft^3	Same	1000 lb, 100 gal
Sensitizers		5000 lb, 500 gal, 650 ft^3	Same	1000 lb, 100 gal
Other health hazards		5000 lb, 500 gal, 650 ft^3	Same	1000 lb, 100 gal

[a] Quantities in any control area (or one-hour fire-resistant area) in excess of the exempt amounts listed below will require compliance with special requirements in the 1988 Uniform Building Code and Article 80 in the 1988 Uniform Fire Code. Amounts are shown for storage, which is the maximum amount allowed in storage and in use, and for use in closed and open systems. Amounts are given in pounds for solids, gallons for liquids, and cubic feet for gases. Exempt quantities can generally be doubled in a sprinklered building and doubled if stored in approved storage cabinets or safety cans. See Tables 9-A and 9-B in the Uniform Building Code for complete information, including combustible dusts and fibers, explosives, and reactive materials.

which may vary in the degree of severity, depending on quantity and concentration, it is often difficult to determine what protection is needed for safe storage and where best to store a particular chemical. For example, concentrated acetic acid is corrosive to the skin (at concentrations of 80% or greater it is defined as a corrosive liquid by Department of Transportation regulations) and it is also a combustible liquid with a flash point temperature of about 43 °C (109 °F). In contrast, vinegar, which has an acetic acid concentration from 4% to 8%, is considered both nontoxic and noncombustible.

When chemicals have multiple hazards it is important to store them in the most appropriate storage areas, and it may also be necessary to segregate them within that storage area. As another example, sulfuric acid and sodium hydroxide might be stored in a single storage area because they are both corrosive, but they would need to be protected from accidental mixing which could cause a hazardous chemical reaction. Separation, segregation, or isolation are recommended depending upon the severity of hazard, total quantities stored, and the size and break resistance of individual containers.

The material and size of storage containers will affect the need for special storage practices and safety procedures. For example, if containers of flammable and combustible liquids are no larger than 5 gallons in size, there is no requirement for special provision to prevent liquid flow from the storage area into the adjoining building.

Ventilation is needed for chemicals and containers that may release dangerous or damaging quantities of vapors or gases that are flammable, corrosive, irritating, or toxic. Ventilation may also be needed for containers and chemicals that may produce annoying odors.

Doors entering storage areas should be identified. Some NFPA standards require specific identification and use of the hazard signal system described in NFPA standard 704 (Figure 3.1).

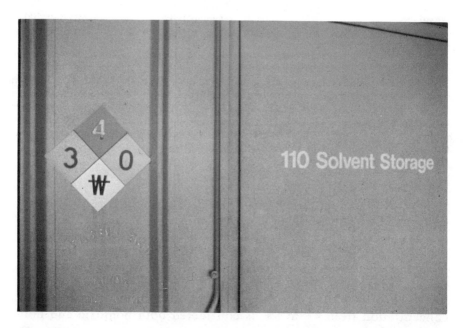

FIGURE 3.1   NFPA 704 hazard signals identifying emergency hazards of chemicals in a storage area.

For every storage area there should be evacuation and emergency procedures to be followed in case of a leak, spill, or fire within the room.

The location of frequently used chemical storage areas should be based on consideration of safety in travel to the storage area and transport of chemicals to and from storage.

The cost of specialized storage space and the cost of time to use the chemical storage facilities are basic considerations. Generally, the cost of space decreases as it becomes less specialized and located farther from the point of use. However, the time cost of getting chemicals increases as the travel distance increases. Chemicals stored at the bench or other work area should be those that are used frequently. Quantities should be limited to the minimum necessary, and the container size should be the minimum that is convenient.

Chemicals stored near the work area in cupboards, cabinets, and closets as backup supplies for the work should be in containers of convenient size, and quantities should be limited to the minimum amounts necessary.

Since storage of volatile or odorous chemicals in laboratory hoods is not economical and does not facilitate safe use of the hoods, we recommend that special ventilated cabinets be provided for storage of such chemicals (Figure 3.2). Cabinets can be ventilated more effectively and at less cost than a laboratory hood. Cabinets should be corrosion-resistant and be shallow enough for convenient use.

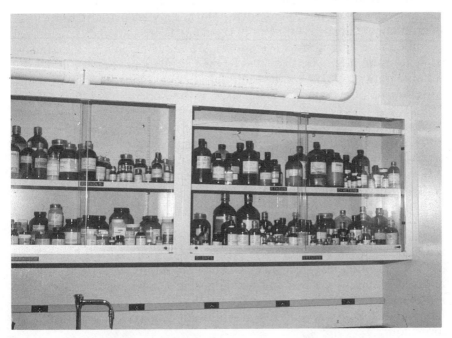

FIGURE 3.2 Ventilated storage for volatile chemicals.

Chemicals in stockrooms or similar accessible supply areas should include all needed chemicals that are not stored in or near work areas. Quantities should be limited to the storage space provided.

Chemicals stored in separate or detached areas should be limited to bulk quantities that cannot safely or economically be stored inside the building or stockroom areas.

Important safety and health considerations should include:

1. Ventilation for protection of health and prevention of corrosion.
2. Lighting for finding containers and reading labels.
3. Identification of storage locations.
4. Strength, stability, and corrosion resistance of shelving.
5. Aisles and storage arrangements that provide for safe access (this may require shallow shelves or pull-out trays or drawers, and movable steps to reach storage above eye level).

3.2.1   Identification of Chemicals

Hazardous chemicals can be stored and handled more safely if they have labels that list precautions or that can be used to refer easily and quickly to safety information (Figure 3.3). Most chemical manufacturers now provide containers of hazardous chemicals with precautionary labeling based on the American National Standard for Hazardous Industrial Chemicals—Precautionary Labeling (ANSI Z 129.1). Adequate precautionary labeling has become more common as a result of the requirements of the Hazard Communication Standard adopted by the Occupational Safety and Health Administration (OSHA), and regulations adopted by a number of states. An example of a labeling system used by one company is described in Chapter 5.

Some labeling terms commonly used are *combustible, flammable, corrosive, irritating,* and *toxic.* Use of a single term will not always provide adequate information because many chemicals have more than one hazard. Another problem is that the term *corrosive* includes materials that may be incompatible with each other. For example, strong mineral acids as well as strong alkaline materials are both corrosive but if they mixed accidentally in a storage area they would react vigorously.

There are several systems and standards for labeling chemicals to communicate their hazards, but there exists no uniformly accepted system for signaling hazards *and* conveying precautionary information. Each standard and system has advantages and limitations, and these are described in the following sections. We believe that a combination of systems is needed so that labeling of laboratory reagents provides adequate safety and health information.

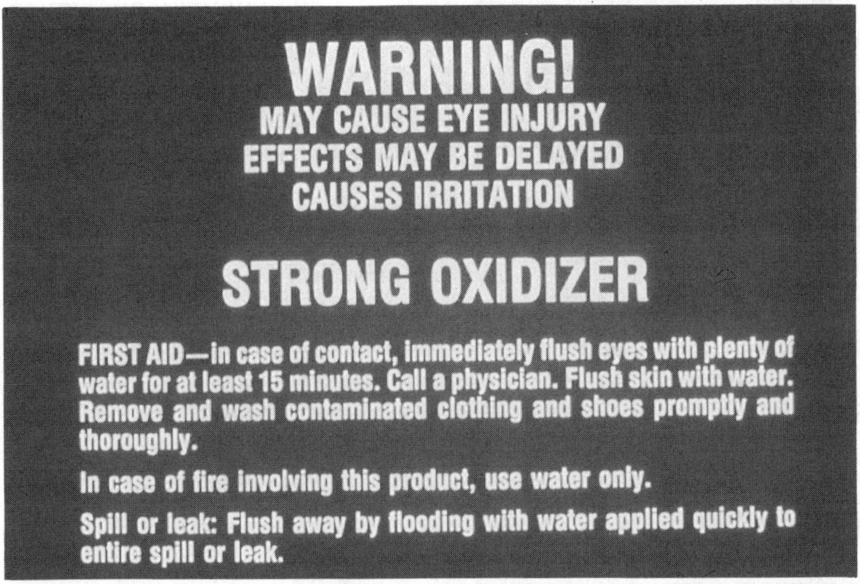

WARNING!
MAY CAUSE EYE INJURY
EFFECTS MAY BE DELAYED
CAUSES IRRITATION

STRONG OXIDIZER

FIRST AID—In case of contact, immediately flush eyes with plenty of water for at least 15 minutes. Call a physician. Flush skin with water. Remove and wash contaminated clothing and shoes promptly and thoroughly.

In case of fire involving this product, use water only.

Spill or leak: Flush away by flooding with water applied quickly to entire spill or leak.

FIGURE 3.3 Example of typical wording on a precautionary label. Note the safety precautions that are related to storage.

3.2.1.1 Precautionary Labeling Many laboratory reagent containers are labeled by the manufacturer with the kind of precautionary information required for consumer products. Precautionary labeling for hazardous materials has been developed by the Chemical Manufacturers Association and the American Conference on Chemical Labeling and adopted as American National Standards by the American National Standards Institute (ANSI). Such precautionary labeling contains the name of the chemical, a signal word such as WARNING or DANGER, the key hazards such as flammable or vapor harmful, and statements of precautions to avoid the hazards.

We believe that precautionary labeling is the minimum that should be required. For containers that are breakable and large enough to create a serious problem if hazardous contents are splashed or spilled, we recommend use of the NFPA hazard signal system to supplement precautionary labeling (Figure 3.4). Use of the hazard signal system is useful for emphasizing emergency hazards.

The U.S. federal government, many states, and some cities have adopted laws and regulations on hazard communication and employee and community right to know. These generally require precautionary labeling on all containers of hazardous materials and that employers have Material Safety Data Sheets available on all hazardous materials in use or storage and provide employees with training to recognize the hazards explained on the labels and in the data sheets.

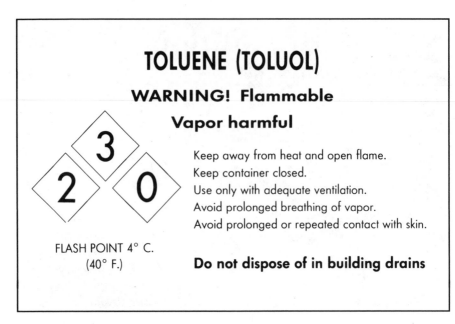

FIGURE 3.4   Example of NFPA 704 hazard signals on a precautionary label.

3.2.1.2 *DOT Hazard Labeling System*   The Department of Transportation (DOT) hazard labeling system uses a color-coded diamond in which there is a symbol (such as a flame) and a term describing the major hazard of the material (Figure 3.5). Hazard classifications include flammable liquid, flammable gas, compressed gas, corrosive, poison, radioactive, explosive, and oxidizer. Most chemicals are rated by what DOT considers the major single hazard, although many chemicals that have hazards in several categories are being required to be labeled and identified with multiple hazard labels.

Because many chemicals have hazards of different types and degrees, we consider the DOT system by itself to be inadequate to communicate the storage and handling precautions that may be necessary for laboratory chemicals.

The DOT regulations for labeling of hazardous materials in the transportation system must be used by laboratories that are packaging and shipping hazardous chemicals as samples, supplies, or waste. In brief, the system requires specification containers and packaging, preparation of shipping papers which list hazardous materials first and in special terms, hazard labels on shipping containers and vehicles, and training of all employees who package and offer hazardous materials for shipment.

3.2.1.3 *NFPA 704 Hazard Signal System*   The NFPA 704 hazard signal system, an NFPA standard, uses a color-coded diamond with four quadrants in which numbers are used in the upper three quadrants to signal the degree of

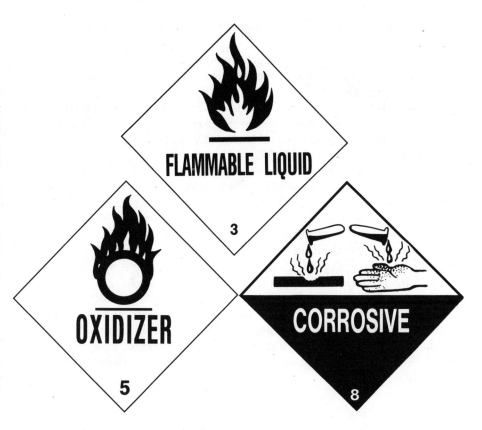

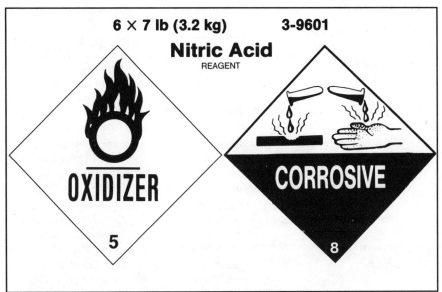

FIGURE 3.5 Examples of DOT hazard labels. Some shipments may have two labels.

emergency health hazard, fire hazard, and instability/reactivity hazard. (The bottom quadrant can be used to indicate water reactivity, radioactivity, biohazards, or any other hazard.)

The three types of emergency hazards are signaled on a numerical scale of 4 to 0, with 4 = extreme hazard, 3 = severe hazard, 2 = moderate hazard, 1 = minor hazard, and 0 = no unusual hazard. Examples of the types of hazards and ratings are shown in Table 3.2.

The NFPA hazard signal system can be useful in alerting personnel to the degree of hazard of the chemical in a container and helpful in drawing attention to some of the storage needs and emergency equipment needed in case of spills or splashes. The limitations of the system are that chronic hazards are not signaled and precautionary information is not given.

As an example of the need for precautionary labeling in addition to hazard signals, the emergency health hazard ratings are estimations of acute exposures by inhalation or skin contact and do not assess chronic exposures or hazards from ingestion or injection.

Hazard signal ratings are useful in assessing the need for precautions for storage, spills, splashes, and other emergency conditions involving major quantities of the concentrated material. Small quantities and diluted materials will usually present less health hazard, less flammability, and less instability.

TABLE 3.2   NFPA 704 Hazard Signal System: Hazard Ratings and Examples

No.	Emergency Health Hazard	Fire Hazard	Instability/Reactivity
4	*Extremely hazardous*	*Flammable gas or class IA liquid*	*Extremely shock sensitive and capable of detonation*
	Hydrogen cyanide	Hydrogen, methane, ethyl ether, pentane	Picric acid, dry; benzoyl peroxide
3	*Toxic or corrosive*	*Flammable liquid, class IB and IC*	*Shock-sensitive materials which may detonate under some conditions*
	Sodium cyanide, sulfuric acid	Acetone, methanol ethanol, toluene	Dilauroyl peroxide
2	*Moderately toxic*	*Combustible liquid class II and IIIA*	*Unstable and water-reactive materials*
	Toluene, ether	Acetophenone	Sopdium, sulfuric acid
1	*Irritating*	*Combustible, including class IIIB liquids*	*Materials that may become unstable under heat or pressure*
	Acetone, MEK	Cod liver oil	Glacial acetic acid
0	*No unusual hazard*	*Noncombustible*	*Not reactive*

TABLE 3.3   Examples of Hazard Signals for Several Laboratory Reagents

| Chemical | Emergency Hazard Signal | | | |
	Health	Fire	Instability	
Acetic acid	2	2	1	
Acetone	1	3	0	
Ethyl alcohol	0	3	0	
Ethyl ether	1	4	1	
Hydrochloric acid	3	0	0	
Hydrogen cyanide	4	4	2	
Picric acid	1	4	4	
Sodium hydroxide	3	0	0	
Sulfuric acid	3	0	2	+ Water reactivity
Toluene	2	3	0	
Xylenes	2	3	0	

Several NFPA codes call for identification of specific storage areas by means of the diamond-shaped, four-quadrant, hazard signal system described in NFPA 704. This system uses numbers ranging from 0 to 4 to indicate the relative degree of hazard in each of three types of hazard: health hazard from short-term contact with or inhalation of the concentrated material, fire hazard, and instability hazard. The definitions of the degrees of hazard are shown in Table 3.2, and examples of the hazard ratings of several common chemicals are shown in Table 3.3.

The hazard ratings that generally should be used in the hazard signal on a storage area are based on the hazard ratings of the material stored there. For example, the combined signal consists of the highest emergency health hazard rating, the highest fire hazard rating, and the highest instability/reactivity rating. However, the NFPA 704 standard does allow reduction of the hazard ratings on the storage area if the hazard rating would be misleading, such as when materials with high hazard ratings are stored in small quantities or dilute concentrations.

It is important that the 704 system be used with judgment so that fire fighters and other emergency response personnel do not become overly apprehensive about hazards that may be relatively minor. For example the hazards presented by certain hazardous chemicals in a laboratory or small stockroom are likely to be much less than the same chemicals in a storage room because of the differences in quantities.

3.2.2   Separation and Isolation

Prevention of fire and other hazardous reactions in storage areas is based partly on avoiding unintentional mixing of chemicals from leaking or broken containers. The need for separation, segregation, or isolation will depend on the size and break resistance of storage containers, the potential for leakage, and the hazards of the chemicals.

Separation is defined by NFPA 49 as storage within the same fire area but separated by as much space as practicable or by intervening storage from incompatible materials. For example, NFPA 49 recommends that sulfuric acid be stored separately from combustible materials, and that acetic acid be stored separately from oxidizing materials (Figure 3.6).

Segregated storage is generally defined by NFPA standards as storage in the same room but physically separated by space from incompatible materials. NFPA 43A requires sills, curbs, or intervening storage to maintain spacing. NFPA 43C specifies separation by at least 20 ft (6.1 m).

Isolation is defined by NFPA 49 as storage away from incompatible materials in a different storage room or in a separate and detached building located at a safe distance. As one example, NFPA recommends that storage of dry benzoyl peroxide (in industrial quantities) be isolated in well-detached, fire-resistive, cool, and well-ventilated buildings with no other materials stored therein.

The 1988 UFC also contains segregation requirements for hazardous chemicals.

FIGURE 3.6   Some chemicals, such as sulfuric acid and acetic acid, should be separated to comply with storage standards. If quantities are limited, as shown here, separation can be achieved by protecting the glass bottle of one of the incompatibles with a bottle carrier, or by a plastic-coated bottle that will resist breakage.

3.2.3 Storage Containers

Large glass containers are susceptible to breakage in transportation and handling unless they are protected by shipping containers, carrying containers, or a heavy plastic coating (Figure 3.7).

Five-gallon metal containers are often too heavy and awkward for convenient and safe pouring of the contents. Fifty-five-gallon drums of solvents are not allowed in laboratory work areas, according to NFPA 45-1986, but only in special rooms equipped for storage or for dispensing.

3.2.4 Ventilation

Ventilation is a fire-prevention requirement for storage rooms in which flammable liquids are stored or dispensed, and it is generally an occupational health recommendation for rooms in which such liquids are dispensed. Even though ventilation may not be required by a code, ventilation is certainly needed for any storage area in which there may be leakage of vapors or gases that are corrosive to metal or irritating, annoying, or toxic to personnel.

FIGURE 3.7 Bottle carrier shown on the left can be fitted over a bottle to prevent breakage and limit spills. Safety can shown on the right provides resistance to breakage and spills, and will open in case of fire to vent vapors and prevent an explosion.

3.2.5 Access and Egress

3.2.5.1 *Building Access* Walkways, ramps, and corridors should be designed to accommodate convenient access. Access for the handicapped is generally required by law, but there are other benefits that justify such designs. For example, any pathway that will accommodate movement of wheelchairs will facilitate movement of pedestrian traffic and delivery carts. Equipment that has to be moved in or out of a building can be moved more easily and safely if there is a well-designed ramp.

Ramps into buildings should be designed to meet optimum standards for access of the mobility-handicapped, with adequately spaced, level rest areas, adequately-sized turning areas, and a gradual slope not exceeding a change in elevation of 1 ft per 12 ft of horizontal run. Provision should be made for heating exposed ramps where ice or snow are likely to accumulate, because any degree of slope can be hazardous if it is slippery.

3.2.5.2 *Movement within a Laboratory Building* Movement within a laboratory building is facilitated by having storage space for movable carts, deliveries, supplies, glassware awaiting use or washing, and waste materials to be picked up. Traffic will move more smoothly if corridors are designed to avoid obstructions by safety equipment and are wide enough to accommodate the amount of pedestrian traffic that can occur. Storage space should be provided for equipment that will be used only periodically.

The minimum width recommended for corridors to allow two people to pass each other is 5 ft (60 in). Design should provide easily maintainable access to emergency devices such as eyewash and safety showers and fire extinguishers. Design should also provide space for respiratory protective equipment in areas not likely to be involved in chemical spills, leaks, or fires.

3.2.5.3 *Egress for Normal Operations and Emergencies* Corridors, stairwells, and other egress components should be increased in width beyond code requirements if two-way pedestrian traffic is expected to be heavy (as in a teaching laboratory) or if corridors may be obstructed by carts or temporary storage. Code requirements for means of egress are minimal because they are designed only for rapid evacuation of all occupants of a building in case of fire.

3.2.5.4 *Location of Exit Access Relative to Hazards* If a storage room has only one exit door, storage should be arranged so that escape is convenient and so that the greatest hazard is farthest from the door. It should not be necessary to travel toward an area of high hazards to escape unless the path of travel is effectively shielded from the hazards by suitable partitions or other physical barriers (Figure 3.8).

The NFPA Life Safety Code defines high hazard occupancy as those buildings having high hazard materials, processes, or contents that are "liable

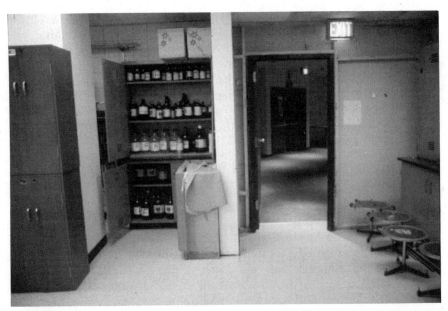

FIGURE 3.8   Hazardous chemicals should *not* be stored where a spill could block the exit from the area.

to burn with extreme rapidity or from which poisonous fumes or explosions are to be feared in the event of fire." In contrast, "ordinary hazard" contents are defined as those that are liable to burn with moderate rapidity and give off a considerable volume of smoke but from which neither poisonous fumes nor explosions are to be feared in case of fire.

3.2.5.5 *Number of Exits from a Storage Area*   Because chemical storage rooms are likely to be a high hazard occupancy, two exit doors should be provided if the room is large enough to make it possible. Two means of egress are generally required by NFPA code as a minimum for every building area of a size that the reasonable safety of occupants may be endangered by the blocking of a single means of egress due to fire or smoke. The means of egress should be remote from each other and arranged so as to minimize any possibility that both may be blocked by a fire or other emergency condition (NFPA 101).

The 1988 Uniform Building Code requires educational laboratories using hazardous materials and H1 and H2 occupancies to have two exits from any room larger than 200 ft^2.

If a second exit is provided from a chemical storage room, it should be checked regularly (such as daily or weekly) to see that it is not blocked on either side and that the door does not become corroded so that it will not work in an emergency.

3.2.5.6   *Exit Access Doors and Direction of Swing*   A door from a room to

an exit or to a way of exit access must be of the side-hinged, swinging type. (Vertical-rolling doors and horizontal-sliding doors may be used to provide a fire-resistant barrier in some locations, but they are *not* permitted in exit routes.) When a room is used for a high hazard occupancy, the door or doors of the room must swing out in the direction of exit travel (NFPA 101). The required exit doors of all laboratory work areas within class A or class B laboratory units must swing in the direction of exit travel (NFPA 45).

3.2.6   Evacuation and Emergency Procedures

Emergencies that can occur within the storage area or that may affect the storage area include fire, spill, release of radioactive material, release of gas, ventilation failure, escape of pathogens, and explosion. Response to such emergencies will usually be limited to primary emergency procedures, unless the facility has provided additional special training for all personnel or has organized and equipped an emergency response team. Emergency procedures will be described in more detail later in this chapter, and requirements for emergency teams will be addressed in another chapter.

3.3   STORAGE OF HAZARDOUS CHEMICALS

The 1988 editions of the Uniform Building Code and the Uniform Fire Code include a great many specific requirements for storage and dispensing of hazardous materials, including spill control, drainage, containment, ventilation, and fire protection. In addition, the UFC requires a storage plan for all storage facilities, including the location and dimensions of aisles and the intended storage arrangement. For storage of toxic compressed gases or highly toxic gases, liquids, or solids the UFC requires that a minimum of two self-contained breathing apparatuses be provided nearby in a safe area. See Table 3.1 for quantities of hazardous materials exempt by the UFC. Greater quantities require special storage (Figure 3.9).

3.3.1   Combustible and Flammable Materials

Many organic and inorganic materials are combustible, and some have such a high degree of combustibility that they are designated as flammable. The most common requirement for special storage is for organic liquids that can release flammable concentrations of vapors at temperatures at or below 93.4 °C (200 °F).

   Since code requirements for special storage are often stated in terms of the fire hazard classification of a material or in terms with a specific degree of hazard it is necessary to describe the classifications and definitions in terms of flash point temperatures.

Classification	Term	Flash Point Temperature
Class III	Combustible liquid	Any flash point at or above 60 °C
Class IIIB	Combustible liquid	At or above 93.4 °C (200 °F)
Class IIIA	Combustible liquid	Below 93.4 °C (200 °F) and at or above 60 °C (140 °F)
Class II	Combustible liquid	Below 60 °C (140 °F) and at or above 37.8 °C (100 °F)
Class I	Flammable liquid	Below 37.8 °C (100 °F)
Class IC	Flammable liquid	Below 37.8 °C (100 °F) and at or above 22.8 °C (73 °F)
Class IB	Flammable liquid	Below 22.8 °C (73 °F) and with a boiling point at or above 37.8 °C (100 °F)
Class IA	Flammable liquid	Below 22.8 °C (73 °F) and with a boiling point below 37.8 °C (100 °F)

Ignitible liquids, regulated by the Environmental Protection Agency under the Resource Conservation and Recovery Act, include Class I flammable liquids and Class II combustible liquids.

Special storage is commonly required for quantities in excess of 120 gallons of flammable liquids and certain combustible liquids that cannot be stored in

FIGURE 3.9 Conveniently usable storage for hazardous chemicals. This provides as much fire resistance as a flammable liquids storage room.

special wooden or metal storage cabinets. The liquids that require such special storage have flash point temperatures at or below 93.4 °C (200 °F), and they include all liquids identified as a flammable liquid, an ignitible liquid, or a combustible liquid in Class II or Class IIIA.

At least one approved liquid storage room is required within any health care facility regularly maintaining a reserve storage capacity in excess of 300 gal (1135.5 L).

Generally, the only other combustible materials that require special storage are combustible gases and a limited number of materials that are classified as flammable solids. Storage requirements for combustible gases will be discussed later in the section on Compressed Gases.

NFPA 30 permits storage for industrial and educational laboratory work to comply with NFPA 45, Standard on Fire Protection for Laboratories Using Chemicals. NFPA 45 establishes limits for the quantities of flammable and combustible liquids permitted to be located in laboratory units, depending on the construction and fire protection provided.

The basic quantities of flammable and combustible liquids permitted in open or unprotected storage can be doubled if they are stored in safety cans or in storage cabinets. Quantities can also be doubled if the area is provided with an automatic fire extinguishing system. Quantities can range from one gallon per 100 square feet in a low-hazard educational laboratory up to 40 gallons per 100 square feet or a maximum of 1200 gallons in a sprinklered high-hazard laboratory unit.

Storage of flammable and combustible liquids outside of flammable liquid storage rooms can be in safety cans or in storage cabinets; there is no requirement that such liquids must be in both safety cans and storage cabinets at the same time.

Flammable liquids and combustible liquids can safely be grouped in a storage room or a storage cabinet that meets the requirements listed above for storage of flammable liquids.

Organic acids are combustible materials and many of them are combustible liquids. Organic acids can safely be stored with flammable and combustible liquids, but they should generally not be stored with mineral acids, which are oxidizing and could react more or less violently with organic acids. Some organic acids that are commonly used in laboratories include:

Acetic acid	Combustible liquid, Class II
Butyric acid	Combustible liquid, Class IIIA
Chloroacetic acid	Usually crystals; flash point 259 °F
Crotonic acid	Combustible liquid, Class IIIA
Oleic acid	Combustible liquid, Class IIIB
Oxalic acid	Organic acid, which is corrosive to skin and highly toxic; DOT label is poison
Stearic acid	Combustible liquid, Class IIIB
Toluenesulfonic acid	Combustible; DOT label is corrosive

3.3.1.1 Flammable and Combustible Liquid Storage Rooms

The requirements for storage of flammable and combustible liquids are spelled out in OSHA standards, in the NFPA Flammable and Combustible Liquids Code, NFPA 30, and in the UFC. The standards apply to storage of flammable liquids, defined as having flash point temperatures at or below 37.8°C (100°F), and combustible liquids, defined for these standards as those having flash point temperatures at or below 93.4°C (100°F), or Class II and Class IIIA.

Both sets of standards allow storage of significant quantities of such liquids within laboratory buildings in rooms that are specially separated from the rest of the building so that a spill or fire in the room is not likely to spread into the main building. See Figure 3.10 for an excellent arrangement of rooms.

An inside storage room for flammable and combustible liquids will be reasonably safe and will meet or exceed both sets of standards if it meets the following requirements. Dispensing will be discussed later in this chapter. An inside storage room that does not exceed 150 square feet in floor area is permitted to contain up to 2 gallons per square foot of floor area in the room, if the room is separated from the building by construction having at least one-hour fire resistance and all openings between the room and the building are protected by assemblies having a fire resistance rating of one hour. If it is desirable to increase the allowable storage capacity of such a room, the capacity can be increased to 5 gallons per square foot by providing the room with an automatic

FIGURE 3.10 Separate rooms for storage of flammable liquids, dispensing of flammable liquids, and storage of flammable waste.

fire extinguishing system, which might be as simple as adding one or two automatic sprinkler heads.

An inside storage room needs to be ventilated to prevent possible accumulation of flammable concentration of vapors from container leaks or spills. Recommended ventilation is from floor level with a capacity of one cubic foot per minute of exhaust for each square foot of floor area in the room, with a minimum of 150 cubic feet per minute. (If there is dispensing in the room, there should be provision for ventilating the dispensing operations close to the points at which vapors are being emitted.)

If the containers of flammable and combustible liquids are no larger than 5 gallons in size, it should *not* be necessary to provide barriers to prevent spills in the room from spreading into the main building. If containers of Class I or II liquids are larger than 5 gallons in size, there is need for curbs, ramps, scuppers, or special drains, with a drainage capacity for all of the water that could be discharged from an automatic fire extinguishing system and from the fire hose streams that may be applied by the fire department.

Drainage from storage rooms for incompatible materials must be kept separate so that incompatible materials do not enter the same system.

Wiring and electrical fixtures located in inside storage rooms must be suitable for the hazards. Explosion-proof (Class I, Division 2) electrical equipment is required to prevent explosions if flammable liquids (Class I) are being stored or dispensed. If only combustible liquids are being stored or dispensed, general use wiring is acceptable.

If an inside storage room has an exterior wall, it will be classified by the NFPA Flammable and Combustible Liquids Code as a "cut-off room," for which there are two additional requirements. First, exterior walls are required to provide ready accessibility for fire fighting. Second, if Class IA or IB liquids are dispensed or if Class IA liquids are stored in containers larger than 1 gallon, the exterior wall or roof is required by NFPA 30 to be designed to provide explosion venting to meet the requirements of NFPA 68.

"Explosion venting" is actually intended to vent a deflagration at relatively low pressures so that personnel are not injured and there is no equipment or building damage. Walls that relieve the pressures that build up in a deflagration merely fall off; they are not propelled as they might be if the pressures build up.

The 1988 UFC contains detailed requirements for explosion venting and suppression.

We do not believe there is a need for explosion venting or deflagration venting in a small room used only for storage. In a room used for pouring or dispensing from a few drums, we do not believe there is need for explosion venting if adequate ventilation is provided to exhaust vapors generated by the pouring or dispensing.

3.3.1.2 *Flammable Liquids Storage Cabinets*   Because hazardous chemicals are used routinely in laboratories, they must be conveniently avail-

able. This requires a certain amount of chemical storage within the laboratories and ready access to frequently used chemicals. Since many laboratory organizations do not wish to have any flammable chemicals on open shelves, there is need for storage within and close to laboratory work areas. Whereas any type of laboratory cabinets will provide some limited fire protection for flammable materials, the minimum that is recognized by code authorities is a flammable liquids storage cabinet with at least 10 minutes protection for the contents.

Protection provided by cabinet construction. Storage cabinets built to meet the specifications of NFPA 30 are designed to insulate their contents so that in case of a fire outside the cabinet the internal temperature will not exceed 325°F for 10 minutes.

Double-walled metal storage cabinets are available in the construction that is deemed to provide the minimum protection required by OSHA and NFPA standards, and some have been tested at Factory Mutual Engineering Laboratories. Most of the commercially available metal storage cabinets have a 60-gallon capacity.

Storage cabinets constructed of one-inch exterior grade plywood are also presumed to meet the fire exposure test described in NFPA and OSHA standards. Fire tests by the Los Angeles Fire Department have shown that the thermal insulation provided by one-inch plywood greatly exceeds that provided by double-walled metal construction (Figure 3.11).

FIGURE 3.11   A storage cabinet made of exterior-grade plywood is acceptable for storing flammable and combustible liquids according to OSHA and NFPA standards.

Plywood storage cabinets have the advantages of providing greater fire protection and being readily fabricated to meet varying quantity needs and to fit conveniently within a new or existing laboratory. Plywood storage cabinets are commercially available, and one brand has passed fire tests and is listed by Factory Mutual Engineering Laboratories.

Specially made storage cabinets that have fire resistance ratings of 30 minutes or longer have become available recently. Records protection cabinets with fire resistance ratings of 60 minutes or longer can be used if there is need for extended fire protection of unique materials.

Ventilation of storage cabinets. Ventilation of storage cabinets is not required by NFPA standards. If it is considered desirable or necessary to ventilate a storage cabinet to control odors or corrosion or to prevent flammable concentrations of vapors, there should be a mechanical system to provide an effective exhaust of the cabinet and the duct should be as fire-resistant as the cabinet. Ducting of plastic, aluminum, or copper with sweated joints is not likely to provide 10 minutes of fire resistance, equal to what a cabinet is expected to provide, and a ventilating duct may reduce the insulation provided for the contents.

Closers on cabinet doors. The UFC requires cabinet doors to be self-closing and provided with a latch. However, automatic closers can be hazardous if they act so rapidly or forcefully that they interfere with safe use of the cabinet.

Storage cabinets are permitted to contain no more than 60 gallons of Class I and Class II liquids, nor more than 120 gallons of Class IIIA liquids. Except in industrial occupancies, no more than three storage cabinets may be located in a single fire area. In industrial occupancies, additional cabinets may be located in the same fire area if individual cabinets or groups of no more than three cabinets are separated by at least 100 feet.

Limits on the number of storage cabinets are, in the opinion of the author, intended to limit the total quantity within the area, so that a greater number of cabinets should be permitted if the total quantity limit is not exceeded. For example, providing several small cabinets at points of usage within a laboratory could reduce the hazards and the time of transporting solvents from a large remote cabinet, without exceeding the total quantity allowed in the area.

3.3.2 Oxidizers

Oxidizers must be stored to avoid contact with incompatible materials such as flammable and combustible liquids, greases, ordinary combustibles, and other materials that could react with the oxidizer or catalyze its decomposition, including other oxidizers.

Mineral acids, including those recognized as strong oxidizers, such as nitric acid, perchloric acid, and sulfuric acid, should be separated from flammable and combustible materials. Such mineral acids should be stored in separate rooms, separate cabinets, or break-resistant containers if large glass bottles have to be stored in proximity to combustible materials. To prevent oxidization of wooden storage shelves (or corrosion of metal shelves), acid-resistant trays or mats should be provided under bottles of nitric, perchloric, and sulfuric acids.

Oxidizing material and four classes of oxidizer are defined by the NFPA Code for Storage of Liquid and Solid Oxidizing Materials, NFPA 43A-1980, which establishes requirements based on quantities. Oxidizing material is defined as any solid or liquid that readily yields oxygen or other oxidizing gas or that readily reacts to oxidize combustible materials. The four classes of oxidizer are Class 1, Class 2, Class 3, and Class 4.

The primary hazard of Class 1 oxidizers is an increase of the burning rate of combustible material with which it comes in contact. Examples are hydrogen peroxide solutions from 8% to 27.5%; magnesium perchlorate; nitric acid, 70% concentration or less; perchloric acid solutions, less than 60% by weight; and silver nitrate. The standard applies when quantities are stored in excess of 4000 lb (1816 kg).

Class 2 oxidizers moderately increase the burning rate or may cause spontaneous ignition of combustible material with which they come in contact. Examples of Class 2 oxidizers are calcium hypochlorite, 50% or less by weight; chromic acid; hydrogen peroxide, 27.5% to 52% by weight; and sodium peroxide. Class 2 oxidizers are regulated when stored in quantities in excess of 1000 lb (454 kg).

Class 3 oxidizers will cause a severe increase in the burning rate of combustible material with which they come in contact or will undergo vigorous self-sustained decomposition when catalyzed or exposed to heat. Examples are ammonium dichromate; hydrogen peroxide, 52% to 91% by weight; perchloric acid solutions, 60% to 72.5%; and sodium chlorate. Class 3 oxidizers are regulated when stored in quantities in excess of 200 lb (91 kg).

Class 4 oxidizers can undergo an explosive reaction when catalyzed or exposed to heat, shock, or friction. Examples are ammonium perchlorate; ammonium permanganate; hydrogen peroxide, more than 91% by weight; perchloric acid solutions, more than 72.5%; and potassium superoxide. Class 4 oxidizers are regulated when stored in quantities in excess of 10 lb (4.5 kg).

Gaseous oxidizing materials such as chlorine, chlorine trifluoride, fluorine, nitrous oxide, oxygen, and about ten other gaseous oxidizing materials not commonly found in laboratories are regulated by an NFPA standard. The Code for Storage of Gaseous Oxidizing Materials, NFPA 43C-1980, applies to oxidizers in cylinders or other containers with an aggregate capacity in excess of 100 lb (45 kg) when in storage or connected to a manifold system.

Emergency water may be needed in all these storage areas because many of the chemicals are corrosive to human tissue.

If the class and quantity of oxidizer are regulated by NFPA standards, storage of oxidizers must be segregated, cut off, or detached, depending on the class of oxidizer. The required fire-resistance for cutoff storage increases as the class increases. Class 4 oxidizers in regulated amounts are permitted to be stored only in detached or isolated storage. Class 3 oxidizers are permitted to be stored only on the ground floor of a building with no basement. Storage areas for Class 2 and 3 oxidizers must be provided with means to vent combustion products in case of a fire emergency, and storage areas for Class 4 oxidizers must be provided with exhaust ventilation for both fire and spill emergencies.

Most gaseous oxidizing materials are highly reactive and can react vigorously with finely divided metals, organic liquids, and other materials that are readily oxidizable.

3.3.3   Corrosive and Irritating Chemicals

In addition to the oxidizers that are corrosive or irritating, alkalies/bases are corrosive or irritating. Those that are liquid in large glass containers, such as ammonium hydroxide, should be stored in a separate cabinet or separate area. Ventilation will be needed for any materials that are volatile.

Emergency water should be provided for areas in which corrosive or irritating chemicals are stored, and construction should limit spread of any liquid spills.

3.3.4   Toxic Chemicals

Storage of toxic chemicals will frequently require ventilation, emergency water in case of chemical splash, and containers or construction to prevent the spread of any liquid spills.

Toxic chemicals that are acid-sensitive, such as the cyanides and sulfides, should be stored so that they are separated or protected from accidental contact with acids.

Pesticide storage must be located or constructed so that runoff from firefighting operations will not contaminate groundwater, streams, ponds, land, or buildings. Storage areas for pesticides and other highly toxic chemicals should be secured when the storage areas are not supervised by a responsible person, so that such materials are not taken without authorization.

The 1988 UFC contains restrictive requirements for storage of quantities of highly toxic liquids and solids in excess of about one liter or about four pounds. Such requirements include special provisions for exhaust systems, spill control, and emergency response equipment. If the highly toxic materials are volatile, the UFC requires two self-contained breathing apparatuses located where they are not likely to be immediately affected by the release of hazardous concentrations of vapors.

3.3.5   Reactive and Incompatible Chemicals

If chemicals are to be stored that are reactive if exposed to air or water, they can safely be stored in sprinklered areas where sprinkler discharge would serve to prevent rupture of the outer container. Temperature control or refrigeration must be provided as needed for chemicals that deteriorate or react if their temperatures exceed safe limits recommended by the manufacturer or person synthesizing the chemical.

Acid-sensitive materials such as the cyanides and sulfides should be stored in a separate location from acids or protected from contact with acids.

Although it is certainly desirable to store chemicals so that incompatible chemicals cannot accidentally mix and hazardous chemical reactions will not occur under routine and emergency conditions, published codes and guidelines cannot cover every chemical that is synthesized, formulated, or produced in quantity.

Some guidelines for segregation of incompatible chemicals may be found in another chapter in this book; in NFPA 49-1990, Hazardous Chemicals Data, and NFPA 491M-1990, Hazardous Chemical Reactions; and in Coast Guard recommendations.

When chemicals are stored in small or break-resistant containers or when they are stored in diluted concentrations, code requirements for separation do not seem logical or necessary. Code requirements for separation seem appropriate for large, breakable glass containers and for drums and large quantities.

3.3.6 Compressed Gases

Combustible gases that are classified as fuel gases may be stored inside of a building up to a total gas capacity of 2500 cubic feet of acetylene or nonliquefied flammable gas or about 309 lb of propane or 375 lb of butane. If there is more than one storage area within a building, the areas must be separated by a distance of at least 100 feet. The quantity of acetylene or nonliquefied flammable gas in a storage area may be doubled if the storage area is protected with an automatic sprinkler system that will provide a density of at least 0.25 gallons per minute per square foot over an area of at least 3000 square feet. (NFPA 51-1977 provides additional detail.) Figure 3.12 shows an excellent way of securing compressed gas cylinders.

NFPA 50A-1978 establishes requirements for gaseous hydrogen systems having containers with a total content of 400 cubic feet or more. The standard also applies where single systems having a content of less than 400 cubic feet of hydrogen are located less than 5 feet from each other.

The 1988 Uniform Fire Code contains detailed requirements for storage of gases that are oxidizing, flammable, pyrophoric, reactive, or toxic. Other storage recommendations are given in publications of the Compressed Gas Association.

3.3.7 Hazardous Waste

Facilities should be designed to provide adequate space and protection for temporary storage of hazardous waste awaiting removal or treatment. Sepa-

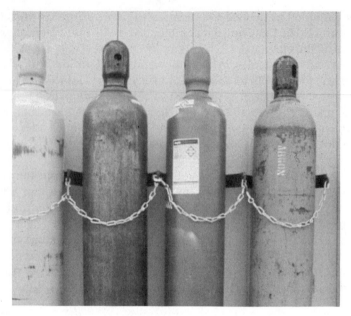

FIGURE 3.12   Compressed gas cylinders should be secured individually to prevent their falling. If a cylinder falls and breaks open it can cause extensive destruction.

rate areas may be needed, depending on the classes of hazardous waste that are to be stored. Storage areas for hazardous waste should be separate from other storage areas for hazardous materials, to prevent damage to new material and to provide better inventory control for all material in storage.

Hazardous waste should be classified into the appropriate hazard class for storage. Hazardous wastes that can be classified into any of the categories for which separate storage is recommended should be protected by break-resistant containers or physical separation. If ventilation is needed to prevent inhalation exposures or corrosion, it should be provided.

Although the EPA does not have any specific standards for the storage of hazardous waste within buildings, the same types of separation and protection should be provided as for hazardous new materials. Fire-protected storage should be provided if quantities of flammable and ignitable wastes are accumulated. Storage areas for waste materials that emit corrosive or toxic vapors should have ventilation, emergency water, and spill limitations.

If hazardous wastes are to be neutralized or otherwise handled before shipment or disposal, it may be advisable to prevent discharge of the wastes from sinks by equipping one or more laboratory sinks with a normally closed valve that is motor controlled to allow users to discharge the sink to the waste line or to a separate outlet for intercepting prohibited discharges.

FIGURE 3.13 Fire extinguishers and other emergency equipment in centralized locations can provide for convenient response to a variety of fire emergencies.

3.4 EMERGENCY EQUIPMENT AND PROCEDURES

Emergency equipment needed for a chemical storage area in a laboratory will usually include a fire alarm system, fire extinguishers, and equipment for responding to chemical splashes, and may include equipment for dealing with chemical spills. (Figure 3.13 shows an example.)

Information necessary for providing fire protection and other hazard control measures will be available if hazards are determined before chemicals are ordered. The Fire Protection Standard for Laboratories Using Chemicals, NFPA 45-1982, spells out the requirement that "When chemicals are ordered, steps shall be taken to determine the hazards and to transmit that information to those who will receive, store, use or dispose of the chemicals."

3.4.1 Fire Emergency Needs, Fire Protection, Blankets, and Extinguishers

3.4.1.1 Fire Alarm A fire alarm system will be required in almost all laboratory facilities, and consideration should be given to installing a system that can be used for other emergency conditions to alert persons in the vicinity and summon emergency assistance. Selection of an alarm system that has multiple uses is recommended. If the alarm system is designed to signal fire evacuation, spill response, or a weather emergency, different signals will be needed. Signage should be provided to describe evacuation or other emergency procedures and to designate evacuation assembly points.

3.4.1.2 *Fire Extinguishers*

Portable fire extinguishers should be installed and located in accordance with NFPA 10, Standard for Portable Fire Extinguisher. Storage areas should have fire extinguishers with ratings appropriate to the hazards, and enough extinguishing capacity for the sizes of fires that may occur.

Almost all flammable and combustible storage areas will require installation of extinguishers with ratings of 20B, 40B, or more to provide adequate fire extinguishing capacity for gallon bottles of flammable solvents. Since no hand-held carbon dioxide fire extinguisher on the market has the extinguishing capacity to put out a one-gallon flammable liquid spill fire, it will be necessary to provide fire extinguishers that contain adequate quantities of dry chemical, halogenated agents, or special foam to have ratings of 20B or greater. Carbon dioxide fire extinguishers are suitable for small solvent fires and are often preferred because they leave no residue.

Multiple-purpose extinguishers are often recommended for areas where fires may involve different classes of materials. For example, an extinguisher with an ABC rating will be the most effective type to use if solvent and paper are both burning and there is live electrical equipment in the same area.

Halogenated extinguishing agents have been preferred for electronic equipment, but extinguishers with such agents will not be rated to extinguish burning paper unless the extinguisher is large enough to have a Class A rating. Special extinguishers with a Class D rating are needed for extinguishing fires in combustible metals such as magnesium and sodium.

Fire extinguishers should generally be located near the doors of storage and work areas, either just inside or just outside the doors. (This location is preferred because an occupant who seeks the extinguisher in case of fire will be heading toward the way out.) NFPA 10 specifies maximum travel distances to fire extinguishers.

NFPA 45 recommends that all laboratory buildings be provided with standpipes and 1½-inch hose for use by occupants to supplement fire extinguishers, and that the hoses be equipped with special nozzles and a water-flow alarm. NFPA also requires that standpipes with 2½-inch hose connections be provided in all unsprinklered laboratory buildings with two or more stories above or below the street floor.

Additional fire extinguishers and fire hoses and protective equipment will probably be required if a laboratory decides to organize an emergency fire brigade on the basis of having special hazards or being in a remote location.

3.4.1.3 *Fire Extinguishing Systems*

Installation of an automatic extinguishing system is recommended to prevent or minimize fire and the resulting disruption. The added expense of such a system is likely to be offset by substantially reduced costs of fire and business interruption insurance.

Automatic fire extinguishment and fire detection systems should be connected to the fire alarm system so that their operation will immediately sound an alarm within the facility. If possible, arrangements should be made to connect internal alarm systems to a central station that will summon the fire department whether the building is vacant or occupied.

3.4.1.4 Fire Blankets   There is no requirement for installation of fire blankets. If fire blankets are installed they should *not* be the vertical type that directs the victim to use it in a standing position, but a type that can easily be taken to the victim. (See Figure 3.14 for a good example.) The recommended emergency response to a clothing fire is to stop, drop, and roll. Standing will increase the amount of hot gases and smoke inhaled.

3.4.2 Emergency Equipment for Spill Response

NFPA 43C, Standard for Storage of Gaseous Oxidizers, requires respiratory protective equipment and other appropriate protective equipment to be readily available outside any storage area in which more than 100 lb of gaseous oxidizing material is stored. (No specific travel distance is stated.)

If spill control material is to be provided, it should be of a type and in a quantity adequate to provide safe control of potential spills. However, as little

FIGURE 3.14   One type of case for a fire blanket which makes it convenient to take the blanket to the victim.

as one gallon of a volatile toxic material such as chloroform *cannot* safely be cleaned up without use of a self-contained breathing apparatus.

If spill control stations are provided, they should be located in a corridor or other area away from those in which spills are likely to occur, so that a spill does not prevent access to the emergency equipment (Figure 3.15).

If the responding fire department will *not* provide service in case of a toxic spill that is nonflammable, there may be a need to provide emergency breathing equipment and protective clothing. If the laboratory must establish self-sufficiency for cleaning up spills of volatile hazardous chemicals, there should be one or more emergency spill stations containing protective clothing, spill control material, and a pair of self-contained breathing masks.

3.4.3 Emergency Water Needs and Requirements

Water is needed in all areas where chemicals are stored so that an adequate amount of water is readily available for emergency flushing of chemicals from the eyes and body. A piped supply of water will be required except in very unusual situations. If shutoff valves are installed in the water line leading to safety drenching equipment, the valves must be the type with an outside stem

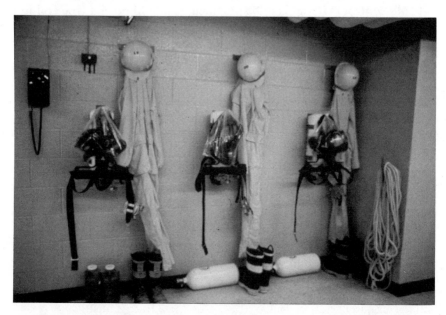

FIGURE 3.15 Essential equipment for a laboratory spill response team. Sets of self-contained breathing apparatus, personal protective clothing, and emergency rope. Equipment is located in an area where spills will not prevent access to or donning of the equipment.

and yoke, and they must be sealed in the open position and labeled for identification. If for some unusual reason a piped water supply cannot be provided, the minimum required is at least a 15-minute supply of water.

Almost all laboratory chemicals and reagents are irritating and many are corrosive or toxic. OSHA standards establish the general requirement and an American national standard defines different types of water delivery systems and sets standards for performance, location, installation, test procedures, and maintenance. This section presents recommendations that supplement the standards, such as on volume of water, drains, water temperature, long-term washing, and decontamination.

3.4.3.1 Recommendations for Emergency Water Devices

Water Supply   The minimum flow rate for emergency showers is 30 gallons per minute, and the minimum flow rate for plumbed eyewash equipment should be at least 3 gallons per minute and preferably 6 to 9 gallons per minute. (The ANSI minimum flow rate for showers is 30 gpm, but only 0.4 gpm for plumbed and self-contained eyewash equipment. We believe that the extremely low flow rate recommended by ANSI for eyewash equipment is suitable only for portable eyewash units made for use in areas without a piped water supply.)

When deciding on the total volume of water needed for flushing, consider that flushing may be required as medical treatment for 30 minutes to many hours, depending on the severity of the burns.

Hand-held drench hoses should generally be installed in combination with all emergency showers to facilitate effective flushing of chemicals from all parts of the body. Since emergency conditions will often require simultaneous operation and use of both a shower and a drench hose, or of a shower and an eyewash unit, the water supply and piping should be capable of providing for simultaneous operation of any combination installed.

Location   Emergency eyewash devices should be located as close as possible to any area where chemicals may be handled. Locating such devices at a sink in the work area will take up less floor area, provide an economical and convenient water supply and drainage, and be easy to find in an emergency.

Emergency showers should be located so that they will not be in the middle of the chemical splash/spill area. If the laboratory space is subdivided into walled modules, emergency showers should be located in the corridors outside the modules. For the purpose of locating emergency showers, we believe that the term *work area* in the OSHA standard should be construed to include the corridors.

The ANSI standard specifies that emergency showers and eyewash devices be accessible within 10 seconds and should be within a travel distance no greater than 100 ft (30.5 m) from the hazard.

If laboratory or storage room personnel are concerned about using a safety shower in a corridor, they might request a shower enclosure. However, a shower enclosure that provides only the minimum unobstructed area recommended by the ANSI standard, 34 inches (86.4 cm) in diameter, will not provide space for the people who will have to assist a splash victim. A small shower enclosure is definitely *not* recommended.

Actuation   Emergency shower and eyewash control valves should be simple to operate and be able to be turned on in one second or less. Valves on emergency water devices should be designed so that the water flow remains on without requiring the use of the operator's hands. These requirements are part of the ANSI standard (Figure 3.16).

Emergency shower operation should be by a cord fastened to a wall so that the shower can be actuated by a person of any height and by someone in a wheelchair (Figure 3.17).

Safety showers should *not* be installed as shown in manufacturers' catalogs, since these usually show the valve directly above the shower head and the actuating handle hanging right next to the head. When a handle in such a location is pulled in an emergency it will be in the way of the victim and the rescuers. The valve should be offset or the actuating handle should be offset so it is out of the way in an emergency. Figure 3.17 shows such an offset.

FIGURE 3.16 One type of emergency eyewash device. The valve is quick-operating and stays on without being held.

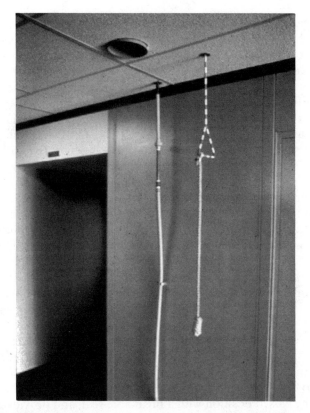

FIGURE 3.17 This safety shower is operated by pulling on the rope. The knotted rope, located by a wall to be easily found in an emergency, is long enough to be reached by a short person or one in a wheelchair. (The hose will provide additional water for localized flushing.)

Floor Drains Since the operation of an emergency shower for the minimum recommended time of 15 minutes at the minimum flow rate of 30 gallons per minute will release 450 gallons of water, floor drains should be installed close to each emergency shower to facilitate the removal of the water that will be released in an emergency. Unless a method can be provided to seal each floor drain with a quickly removable cap, consideration should be given to providing automatic trap fillers or other plumbing arrangements so that traps in the floor drains do not dry out. (It has been suggested that a floor drain could be kept filled by connecting to it the waste line from a drinking fountain.)

Water Temperature Cold water is recommended for emergency flushing of chemicals from the eyes and body, as well as for emergency treatment of thermal burns. Cold water is generally best because it will slow the reaction rate of the chemical splashed, constrict blood vessels and minimize circulation of an absorbed chemical, slow cellular metabolism and enzyme reaction rates, and help reduce the pain of chemical contact. Tepid water may be too warm to minimize adverse effects of splashed chemicals, and hot water can increase injury. However, prolonged washing with cold water can be painful and may on large splashes cause serious loss of body heat.

When prolonged washing may be needed and medical judgment may require water temperatures above those coming from cold water lines, there are several ways in which the water can be tempered. Eyewash devices attached to the outlet of the regular mixing faucets on a laboratory sink provide an economical and reasonably safe way to adjust water temperature for prolonged emergency washing of the eyes and face. Water temperature for indoor showers can be raised above that delivered by a cold water line by automatic tempering valves, by in-line holding tanks, and by manually controlled valves. Automatic water tempering devices can be dangerous if they malfunction and yield hot water, and they are not recommended except in unusual situations.

If in-line tanks are installed within a heated building to provide a volume of water at ambient building temperature, the water will probably need to be recirculated to prevent growth of microorganisms that can seriously contaminate the water.

Safe and economical ways to furnish manually controlled temperatures to emergency showers are to provide each with a normally closed connection to a hot water line, or to provide special emergency shower rooms. (In open areas, the valve to the hot water line for an emergency shower should be accessible, sealed in the closed position with a breakable seal, and provided with a sign that indicates that the valve may be opened in case of emergency to temper the water.)

One or two emergency shower rooms should be located in each laboratory building in which hazardous chemicals are used or stored. The floor of the room should be as slip-resistant as possible, and the room should be large enough to accommodate at least one victim and two or more members of the rescue team. If possible, one or more wooden shower benches should be provided. Consideration should be given to accommodating two or more victims of one emergency (Figure 3.18).

For emergency showers in outdoor locations, such as at a hazardous waste storage area, the water may need to be tempered to keep it from getting dangerously hot or from freezing. In areas where water lines are on or near the surface of the ground and subject to heating by the sun, it will be important to provide water at temperatures that will not cause thermal burns or aggravate chemical burns. Water lines may need to be insulated and water circulated to prevent solar heating.

In areas where water lines may freeze, it will be necessary to provide some means of heating the water supplied. Keeping water lines from freezing may require a carefully controlled system for heating the water directly, or a heated building in which a large emergency water tank can be kept warm indirectly.

FIGURE 3.18 A safety shower for long-term washing of a chemical splash victim. The room has six shower heads supplied with tempered water, and it is large enough to accommodate a victim and one or more rescuers to help the victim remove clothing and wash chemicals off.

3.4.3.2 OSHA Standards for Emergency Water The general OSHA requirement for emergency water is: "Where the eyes or body of any person may be exposed to injurious corrosive materials, suitable facilities for quick drenching or flushing of the eyes and body shall be provided within the work area for immediate emergency use." Another section of the standards includes a guideline for providing emergency water for splashes of liquids that may burn, irritate, or otherwise be harmful to the skin. It calls for a piped supply of clean cold water with a quick-opening valve and at least 48 inches of hose not smaller than three-fourths inch. As an alternative, deluge showers and eye flushes are to be provided where harmful chemicals may be splashed on parts of the body.

3.4.3.3 ANSI Standard for Emergency Eyewash and Shower Equipment The ANSI standard for emergency eyewash and shower equipment, ANSI Z 358.1-1990, defines four different types of water delivery systems and sets standards for their installation, performance, test procedures, and maintenance. (The ANSI standard also sets forth requirements for training.) The four types of water delivery systems are emergency showers, plumbed and self-contained eyewash equipment, eye/face wash equipment and hand-held drench hoses, and combinations of shower and eyewash equipment.

Emergency Showers   The ANSI standard for emergency showers calls for shower heads 82–96 inches above the floor with a spray pattern centered at least 16 inches from any obstruction and 20 inches in diameter at a height of 60 inches above the floor. The valve actuator is to be easily located and readily accessible, with a handle no more than 69 inches (175.26 cm) above the standing level, and the control valve is to remain open without being held. The standard requires that emergency showers be capable of delivering a minimum of 30 gallons per minute (113.6 L/min) of water, and that the shower be connected to a 1-inch I.P.S. minimum water supply.

Plumbed and Self-Contained Eyewash Equipment   The standard specifies that "a means shall be provided to assure that a controlled flow of potable water or its equivalent is provided to both eyes simultaneously at a low enough velocity so as not to be injurious to the user." The specified volume of water is not less than 1.5 liters per minute (0.4 gallons per minute) for a period of 15 minutes.

Eye/Face Wash Equipment and Hand-Held Drench Hoses   The standard for eye/face wash units is similar to that for eyewash units, but the flow rate recommended is 11.4 liters per minute (3.0 gpm). Drench hoses are to be designed to provide a controlled flow of water to the eyes or a portion of the body at low enough velocity to avoid injury to the user, and to deliver water at a minimum rate of 11.4 L/min (3.0 gpm).

Combinations of Shower and Eyewash Equipment   The standard sets similar requirements for combinations of emergency water delivery systems, but states that, "It is not necessary for all components to operate simultaneously (individual conditions will dictate this requirement)."

Water Temperature   The ANSI standard does not specify delivered water temperatures for emergency shower and eyewash units but the appendix of the standard does refer to the need to maintain a temperature that will be safe for the user.

3.4.4   Emergency Procedures

There are four emergency procedures for which all personnel should be trained:

Fire emergencies
Clothing
Spills
Chemical splashes

All these emergencies require action by whoever is in the vicinity and can respond immediately. Trained personnel or an emergency team can continue the emergency procedures after the initial action by personnel in the vicinity.

There are two problems that may complicate an emergency: failure of personnel to respond promptly to the emergency, and failure of personnel to recognize the need to summon additional help.

3.4.4.1 *Primary Emergency Procedures*   Primary emergency procedures include the following steps:

1. Alert personnel in the immediate vicinity of the emergency.
   a. Give the nature and the extent of the emergency.
   b. Give instructions.
      (1) Call the Fire Department.
      (2) Sound the alarms.
      (3) Close doors.
2. Confine the emergency.
   a. Close doors to prevent spread of fire, smoke, vapor, gas, and fumes.
      (1) Close doors to corridor to help confine the emergency to the storeroom.
      (2) Close doors to stairwells to help confine the emergency to one floor.
3. Evacuate the building or section involved.
   a. An evacuation alarm system is needed and generally required.
   b. Evacuation procedures should be posted.
   c. Assembly points should be designated for personnel accounting.
   d. Evacuation and assembly should be practiced in drills.
4. Summon assistance.
   a. Call the fire department.
   b. Give the location and type of emergency.

3.4.4.2 *Clothing Fire Emergency Procedures*

1. Stop the person on fire from running!
   Do not allow anyone to run, not even to a fire blanket.
2. Drop the person to the floor or other horizontal surface.
   Standing will allow flames to spread upward.
   Standing in a fire blanket can funnel hot gases to the eyes and nose.
3. Roll the person to snuff out the flames.
   Blankets can be useful if they are brought to the person.
4. Cool the person.
   Remove smoldering clothing.
   Use water or ice packs to cool burns and minimize injury.
5. Get medical assistance.

3.4.4.3 *Spill Emergency Procedures*   The initial procedures for toxic chemical spills are the same primary procedures as those recommended for fires and other emergencies:

Alert personnel in the vicinity.

Confine the emergency by shutting doors.

Evacuate the emergency area.

Summon assistance.

If the spilled chemical is as toxic and hazardous as chloroform, self-contained breathing apparatus will be *essential* to protect cleanup personnel from concentrations that can be immediately dangerous to life or health.

After cleanup personnel have been protected with necessary protective clothing and breathing apparatus, there are two approaches to dealing with the chemical spill:

1. Use the floor as the reaction vessel for neutralization.
2. Absorb the chemical and carry out the reaction elsewhere.

Several of the spill kits that have been sold use the principal of neutralization on the floor. In some cases, spills may require neutralization in place, particularly if the material has splashed on walls and the ceiling.

Having spill kits will provide a misleading sense of confidence if their capacity is less than that of the quantities that can be spilled. For example, many common spill kits have a rated capacity of only 4 ounces.

Having spill-absorbent material in adequate amounts for possible spills is important, but the next problem is of having ways of distributing the material effectively over the spill, not just in isolated piles.

Bear in mind that when you pick up spilled hazardous material, you have just created hazardous waste that will have to be handled according to the regulations of the Resource Conservation and Recovery Act.

3.4.4.4 *Chemical Splash Emergency Procedures*

The first step in any chemical splash is to get the concentrated chemical off the skin, and the second step is to desorb from the skin as much of the chemical as possible.

Reduction of chemical contact with the skin usually requires removal of contaminated clothing, to remove as much splashed chemical as possible and to prevent chemical on the clothing from being flushed through to the skin. Copious amounts of water are recommended and needed to dilute splashed chemicals.

Large amounts or pieces of water-reactive chemicals should be quickly brushed off the skin, if possible, but emergency water should be used for diluting and cooling so that such chemicals do not react with the moisture in the skin. We know of only one class of compounds that should not be flushed off with emergency water, nitrogen mustard compounds.

Desorbing chemicals from the skin will take approximately ten times as long as the chemical was in contact with the skin. The generally recommended time for washing a splash is 15 minutes, which is a minimum. The time required for adequate flushing may be as long as several hours.

Immediate use of emergency water takes precedence over transporting the injured person to a medical facility.

Prevention of chemical splashes is the best procedure, by use of eye and face protection and protective clothing.

3.5 DISPENSING

Ventilation, spill-limiting construction, and emergency water will usually be required for safe dispensing of chemicals in volume or from shipping containers. If the chemicals are flammable liquids, bonding and grounding will need to be provided and explosion-proof electrical equipment will be required within the hazardous location, as defined by the National Electrical Code, NFPA 70.

Provisions should generally be made to control spills and to limit their spread beyond the dispensing area. Leakage from dispensing containers or overfilled containers can cause damage or hazardous conditions in the dispensing area or in locations to which filled containers are taken.

General dilution ventilation will be needed to prevent accumulation of vapors from leaking containers, and special local exhaust will be needed at the points of dispensing to remove vapors that can cause corrosion, adverse health effects, or accumulation of flammable concentrations of vapors.

New and remodeled laboratories should have ventilated cabinets for storage and dispensing of volatile hazardous chemicals so that laboratory hoods will *not* be used for storage or dispensing.

Emergency equipment and an alarm station should generally be provided to be used in case of spill, splash, or fire. Additional provisions should be provided for personnel protection if anyone is allowed to work alone in a dispensing area; such measures might include monitoring the area for calls for help or monitoring it by television.

Flammable or combustible liquids should be dispensed within a separate room designed solely for dispensing, not in a room designed and used for storage of closed containers. Dispensing such liquids is prohibited by NFPA 30 in flammable liquid storage areas larger than 1000 square feet, and the author does not recommend dispensing in any storage area. Such liquids must not be dispensed in general storage areas unless the dispensing area is suitably cut off from ordinary combustibles and flammable liquid storage areas by fire-resistant construction.

Dispensing chemicals safely may require special equipment and procedures to prevent leakage, vapor dispersion, or fire. Leakage from dispensing containers or overfilled containers can cause damage or hazardous conditions in the dispensing area or in locations to which filled containers are taken. Dispersion of vapors can cause corrosion, adverse health effects, or accumulation of flammable concentrations of vapors. Fire or explosion can be caused by ignition of flammable vapors by portable ignition sources, static electricity, or fixed electrical equipment.

Drums from which liquids are dispensed should sit in a vertical position, if possible, rather than on their sides in cradles. Gravity dispensing of flammable liquids is not allowed unless valves are self-closing (Figure 3.19).

Dispensing from drums in the vertical position will reduce the need for floor space and manual handling, minimize leakage from valves and bungs, and make it more convenient to provide adaptable local ventilation.

If dispensing is done frequently, a compressed air supply can be provided for operating air-driven pumps to transfer the liquids. Air pressure must not be used directly on 55-gallon drums or other shipping containers. Pressurizing a drum or other shipping container with air from a compressed air line is hazardous and prohibited by code.

Dispensing of chemicals may also require special equipment and procedures if personnel are working alone in dispensing areas. Emergency equip-

FIGURE 3.19   Safety can used for gravity dispensing has a self-closing valve.

ment and an alarm system should generally be provided in case of spill, splash, or fire.

Dispensing of flammable and combustible chemicals has special safety needs, many of which are required by codes developed by the National Fire Protection Association and adopted by governmental jurisdictions. For example, no flammable or combustible liquid should be stored or transferred from one vessel to another in any exit corridor or passageway leading to an exit. Class I and II liquids must not be dispensed in general storage areas unless the dispensing area is suitably cut off from other ordinary combustibles or liquid storage areas, and dispensing is not permitted in cutoff rooms or attached buildings larger than 1000 square feet in area.

Whenever possible there should be no dispensing within a flammable liquid storage room, and dispensing and transfer operations should be separated from storage areas containing flammable and combustible liquids. If a fire does occur as a result of dispensing, there will be less fuel and less damage if the fire cannot spread to stored material. However, if there are severe space limitations and storage and dispensing cannot be separated, recommendations for dispensing should be followed in addition to the recommendations for storage.

Article 80 of the 1988 UFC contains some detailed requirements for dispensing hazardous materials other than flammable liquids. It contains ventilation requirements, requirements for explosion venting and suppression, and additional requirements for emergency prevention and response.

3.5.1 Preventing Leakage

There are four important aspects of preventing leakage as a result of dispensing operations:

1. Controlling pouring and gravity dispensing with appropriate valves
2. Limiting any pressure applied to dispensing containers
3. Providing expansion space within all containers that are filled
4. Protecting all large glass bottles from breakage

Dispensing of liquids can be accomplished by pouring, gravity flow, pressure, or pumping.

Pouring safely from a container depends on an individual's ability to lift and hold the container and to control the angle and rate of flow. It is generally safer to provide equipment that will hold the container and allow it to be turned for pouring, and that will return the container to an upright position after pouring.

Dispensing by gravity flow by means of a valve in the bottom of the dispensing container is convenient. However, the valve should not leak and it must be self-closing if it is used for dispensing any combustible or flammable liquid. This requirement is based on preventing uncontrolled flow of the liquids if the valve is left unattended.

Liquids can be transferred by use of a siphon, but the procedure must be carried out so that there is no personal contamination in starting the siphon and no excess flow if the siphon is left unattended. Starting a siphon flowing by pressurizing the container with a hand-squeezed bulb will not generate enough pressure to deform or rupture the drum.

On the other hand, pressurizing a shipping container with air from a compressed air line is hazardous because such containers can easily be over-pressurized so that the air pressure blows liquids out of the container. Use of air pressure is prohibited for transferring any flammable or combustible liquid.

Transferring liquids by pressure of inert gas is permitted by NFPA 30 only if the design pressure of the container is known and if controls and pressure relief devices are provided to limit the pressure so it cannot exceed the design pressure.

Air pressure can safely be used for transferring liquids if the air pressure operates a pump, without pressurizing the container.

Pumping liquids from an opening in the top of the dispensing container avoids the necessity of tipping the container or positioning it horizontally for gravity dispensing and results in safer handling of drums and use of less floor space. Pumping will avoid leakage through a drain valve. One disadvantage is that a pump will be required for each liquid that could be contaminated by dispensing with a pump that is also used for other liquids.

Providing expansion space within all containers that are filled is important because overfilling containers can result in pressures great enough to cause leakage or to rupture the container. Safety cans, which have a spring-loaded lid, will vent vapors if they are filled at a temperature less than that in the area to which they are taken and stored. Glass bottles with screw cap lids can rupture if they are filled nearly to the top with cold liquid and then stored in a warm or hot area.

Protective measures should be taken to prevent breakage of large glass bottles during storage or transportation within the facility. Plastic-coated bottles can be purchased that are suitable for transporting and using liquids that are heavy, corrosive, extremely flammable, extremely toxic, or very expensive. Two-piece plastic jackets are available for protecting some types of gallon reagent bottles (see Figure 3.7). Several types of bottle carriers are also available.

Protection of large glass bottles of chemicals with breakage-resistant carriers or coatings can provide effective separation for containers of incompatible chemicals that could react if mixed.

3.5.2 Preventing Vapor Dispersion

Ventilation is needed for dispensing operations that may disperse vapors or aerosols that are corrosive, irritating, toxic, or flammable. Corrosive vapors or aerosols can damage the storage facility, equipment, and other containers. Corrosive, irritating, or toxic vapors can cause discomfort or adverse health effects. Flammable vapors can accumulate in concentrations which if ignited could result in a flash fire or explosion.

Local exhaust ventilation should be provided that will allow exhausting within a few inches of every point at which hazardous liquids are transferred, because dispensing liquids may disperse vapors or aerosols (Figure 3.20).

The ventilation required by the NFPA Flammable and Combustible Liquid Code is primarily dilution ventilation to prevent accumulation of flammable vapors from leaking storage containers. Ventilation of flammable liquid storage rooms is required for floor areas where flammable vapors can collect.

The ventilation required by the NFPA code is not designed to protect personnel from exposures in dispensing operations. Effective ventilation of dispensing operations will exhaust vapors from the point at which the vapors are dispersed, and away from the breathing zone of the person doing the dispensing. See Figure 3.21 for an outstanding example of good ventilation for dispensing.

The NFPA code recognizes that local or spot ventilation may be needed for control of health hazards, and allows such ventilation to provide up to 75% of the ventilation required by the code for the room. The code requires one cubic foot per minute per square foot of floor area (0.028 m^3 per 0.0929 m^3).

FIGURE 3.20   Ventilation provided at a sink to capture vapors and aerosols generated by pouring and cleaning operations.

FIGURE 3.21 Ventilation provided for dispensing from many containers of flammable or toxic solvents.

The mechanical ventilation system for dispensing areas must, according to the NFPA code, be equipped with an airflow switch that will sound an audible alarm if the ventilation fails.

The NFPA Fire Protection Standard for Laboratories Using Chemicals, NFPA 45-1986, specifies that transfer of Class I liquids to smaller containers from bulk stock containers not exceeding 5 gallon (18.9 L) in capacity inside a laboratory building or laboratory work area shall be made:

1. in a laboratory hood; or
2. in an area provided with ventilation adequate to prevent accumulations of flammable vapor/air mixtures exceeding 25% of the lower flammable limit; or
3. in a separate inside storage area, as described in NFPA 30.

Transfer of Class I liquids from containers of 5 gallon (18.9 L) or more capacity must be carried out:

1. in a separate area outside the building; or
2. in a separate area inside the building in a storage area that meets NFPA requirements.

NFPA 45 explains that ventilation for transfer operations should be provided to prevent overexposure of personnel while transferring flammable liquids, and that control of solvent vapors is most effective if local exhaust ventilation is provided at or close to the point of transfer.

Explosion venting is not required for separate inside storage areas if containers are no greater than 60 gallons (227 L) and if transfer from containers larger than 1 gallon (3.785 L) is by means of approved pumps or other devices drawing through a top opening, according to NFPA 45.

3.5.3 Preventing Ignition of Vapors

The NFPA Flammable and Combustible Liquid Code requires that precautions must be taken to prevent the ignition of flammable vapors, and states specific requirements for controlling ignition sources where flammable or combustible liquids are dispensed. Ignition sources include open flames, smoking material, cutting and welding operations, hot surfaces, radiant heat, frictional heat, static electricity, electrical and mechanical sparks, spontaneous combustion, and heat-producing chemical reactions.

Hot work, such as welding or cutting operations, use of spark-producing power tools, and chipping operations should not be permitted except under supervision of a responsible individual who will make an inspection of the area before work begins to be sure that it is safe for the work to be done and that safety procedures will be followed.

Static electricity is generated when liquids are dispensed, and under some conditions it may accumulate to voltages high enough to cause discharges that can ignite flammable vapors. Class I liquids, and Class II or Class III liquids at temperatures above their flash points, must not be dispensed from a metal container into a metal container unless there is an electrical connection between the containers, by maintaining metallic contact during filling or by a bonding wire between them, or by other conductive path with an electrical resistance not greater than 10 ohms. Figure 3.22 shows a good clamp. Such electrical connection or bonding is not required where a container is filled through a closed system or one of the containers is made of glass or other nonconductive material.

Electrical wiring and equipment located in inside rooms used for storage of Class I liquids must be suitable for Class I, Division 2 classified locations. Such equipment is commonly called explosion-proof. See Figure 3.23 for an example of identification of such specialized equipment. In rooms where dispensing of Class I liquids is permitted, electrical systems must be suitable for Class I, Division 2 classified locations, except that within 3 feet (0.0914 m) of a dispensing nozzle area the electrical system must be suitable for Class I, Division I locations.

The NFPA code specifies that electrical equipment in ventilated areas used for drum or container filling be labeled by the manufacturer as suitable for Class I locations, according to the National Electrical Code. The NFPA codes specify that the electrical equipment within 3 feet in all directions from filling and venting openings must be suitable for Division 1 locations, and that the

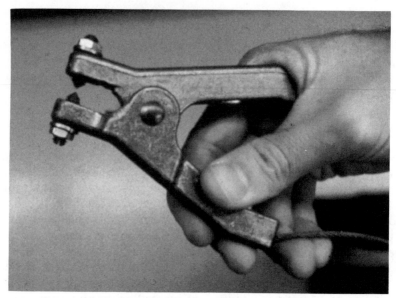

FIGURE 3.22 A clamp for bonding and grounding. This type of clamp will provide excellent electrical contact with a metal container.

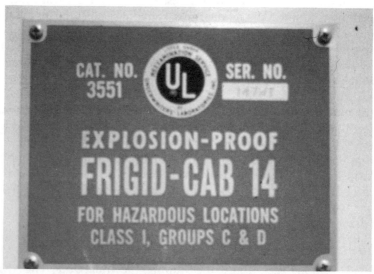

FIGURE 3.23 Example of a refrigerated storage device that can safely be used for storage of volatile flammable liquids.

equipment between 3 and 5 feet from vent or fill openings must be suitable for Division 2 locations. Electrical equipment up to 18 inches above floor or grade level within a horizontal radius of 20 feet from vent or fill openings must also be suitable for Division 2 locations.

3.5.4 Working Alone

Since working alone in a dispensing area can be hazardous in case of accident or overexposure, extra precautions are usually needed. Precautions should be related to factors such as the volume of material dispensed during a given period of time, the hazards of the materials being dispensed, the size of the dispensing area, and whether a person working in the dispensing area can be seen by someone outside the area.

Extra precautions do not seem necessary if the dispensing consists solely of someone going into the area occasionally to fill a small container such as a liter bottle with a material with no unusual hazards.

Extra precautions do seem necessary if filling is done frequently, if one person works for a long time in the area, or if the dispensing area is located remotely or in an area where there is infrequent pedestrian traffic.

Where the interior of a dispensing room cannot be seen from outside the room, it is a common precaution to keep the door open whenever anyone is inside the room.

If it is not feasible to have a second person present during dispensing operations, the hazards may warrant an audible monitoring system that will allow another person to hear immediately any loud noise or call for help.

3.5.5 Emergency Alarms

In addition to the emergency equipment recommended for controlling chemical splashes and spills, each dispensing area should be provided with an emergency alarm station outside the dispensing area. The emergency alarm station could consist of a conventional fire alarm pull station, a special emergency pull station, or an emergency telephone that sounds an alarm and is answered when the telephone is picked up.

3.6 INVENTORY

3.6.1 Inspection of Storage Containers and Facilities

Management of chemical storage should include periodic inspection of storage containers, shelving, and storage facilities. Inspection should determine whether there has been any corrosion, deterioration, or damage as a result of leakage or spills from containers. Containers should also be inspected to ensure that labels are legible and fastened to their containers.

Information has been published on the hazards of prolonged storage of peroxidizable compounds, picric acid, and some other materials that become dangerous or deteriorate over a period of time. However, there is no published information on the length of time that chemicals in unopened containers can safely be stored. How long chemicals can safely be stored unopened will depend partly on the chemical and partly on the environmental conditions in the storage area.

Chemicals that lose their identification because their labels deteriorate or fall off cannot safely be used or disposed of without analysis to determine their identity and hazards.

3.6.2   Dating of Laboratory Reagents

Management of chemical storage should include dating of all containers when received and when opened. Dating of containers of laboratory reagents is usually required so that materials that have a limited shelf life can be replaced to avoid interference with accurate test results. Equally important is the dating and management of materials that can deteriorate and produce severe hazards. For example, picric acid can dry and ethers can peroxidize, creating explosive hazards. It is critical that each container of material that can deteriorate in prolonged storage be dated when it is received for storage and that peroxidizable materials be dated when they are first opened.

Ethers and other peroxidizable compounds kept for prolonged periods after they have been opened will form peroxides that can react explosively when the container cap is removed or when they are concentrated during laboratory activities. In one case, the person removing the stopper of an old bottle of ether was killed when the peroxides in the closure were detonated by the friction of opening the container.

Ethers should be managed safely by the following procedure:*

1. Buy ethers in the smallest amount possible (to limit amounts exposed to air).

2. Date all ethers when they are opened.

3. Test all opened ethers for peroxide concentration within a few months from the date the ethers are opened and at regular intervals thereafter. Test ethyl ether and dioxane every six months during storage and test isopropyl ether every three months.

4. If peroxide concentrations are acceptable, redate the container and retest at the next scheduled test date.

---

*Recommendations for identifying peroxidizable compounds, limiting shelf life, testing for peroxides, and removing them appear in "Control of Peroxidizable Compounds" in *Safety in the Chemical Laboratory*, Vol. 3, Div. of Chemical Education, Publications Coordinator, 215 Kent Rd., Springfield, PA 19064.

5. If peroxide concentrations are not acceptable, remove the peroxides or dispose of the ether as hazardous ignitable waste.

6. If peroxide concentrations cannot safely be determined but it is believed they may be excessive and possibly explosive, plan to have the entire container removed with *extreme* care!

3.6.3 Disposal of Shock-Sensitive Chemicals

If you find shock-sensitive compounds that need to be disposed of, be sure that the disposal team recognizes the hazards and plans the disposal in ways that will not threaten personnel or facilities. It is advisable to schedule the removal of potentially explosive compounds when it will be possible to evacuate the area and when there will be a minimum exposure of personnel (such as late at night).

Ethers with high concentrations of peroxides are extremely sensitive to impact and physical shock and are capable of violent explosion. Ethers with unknown concentrations of peroxides should be handled as if they were booby-trapped. Avoid letting anyone shake the container to see how much is in it. No shock-sensitive compound such as old ethers or old picric acid should be handled with disregard for their explosive power.

3.6.4 Management of Space

When a chemical storage space needs to be used to the maximum, compatible materials should be stored by the size of the container. For example, a lot of potential storage volume is wasted if shelves spaced vertically for storage of gallon bottles are used for storage of small bottles.

The ability to select the appropriate storage location and to find and manage stored chemicals will be greatly improved by establishing a system for identifying and coding both the chemicals and the storage spaces. If laboratory chemicals and their containers can be identified by a unique number (such as the Chemical Abstracts Registry Number), and if the storage locations can be identified by numbers, the chemicals can be stored in a safe and retrievable manner with less confusion and less possibility of losing track of the chemicals in an alphabetical storage arrangement.

For example, one company with a large research organization and many organic chemicals has set up a system in which:

on the top of each small bottle is a unique code number in addition to the chemical name on the side of the bottle.

small bottles are stored in drawers in a ventilated cabinet.

each cabinet and drawer and section has an identifying letter or number (Figure 3.24).

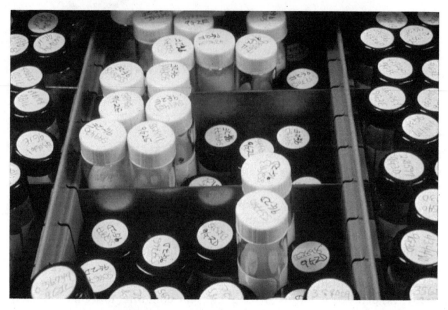

FIGURE 3.24   An example of storage of organic chemicals in a drawer.

Each chemical is locatable by reference first to an alphabetical list that is computer-generated and cross-indexed with commonly used terminology. The reference list gives the unique chemical code number and the coordinates of the storage location. The chemical can be obtained by going then to the storage room, cabinet, drawer, and drawer section. The code number on top of the bottle identifies the chemical, and the name on the bottle confirms that the search has been successful.

Storing a combination of large and small containers on the same shelf can make retrieval difficult and bottle breakage likely. If deep shelves spaced vertically for gallon bottles or 5-gallon cans are used for storage of small containers, storage volume is wasted. To design chemical storage space that can be used to the maximum, compatible materials can be stored by container size and the shelves sized and spaced appropriately.

The storage density of small bottles of chemicals can be increased by storing them in drawers mounted on heavy-duty, full-extension glides that allow convenient access to all parts of the drawer. Drawer storage has been used in at least one major laboratory complex, even though there seems to be a need for a drawer system specifically designed for storage of chemicals.

Once the general storage location of chemicals is determined by compatibility requirements and container sizes, specific storage locations can be identified by numbers and letters. With the "home" of each chemical listed following its name on an alphabetical inventory, it will be easy to find the loca-

TABLE 3.4 A Chemical Inventory and Locator System

Chemical Name	C.A.S. Number	Room Name[a]	Aisle Number	Rack Number	Shelf or Drawer Number	Row Number	Position in Row
Acetic acid	64-19-7	FLAM	1	1	A	1	A
Acetone	67-64-1	FLAM	1	1	A	1	B
Ammonium hydroxide	133502105	ALK	2	2	C	3	A
Anhydrone	See magnesium perchlorate						
Magnesium nitrate	10377-60-3	OXID	4	4	2	7	D
Magnesium perchlorate	10034-81-8	OXID	1	2	B	2	C
o-Methoxyphenol	See guaiacol						
p-Methoxyphenol	150-76-5	FLAM	7	3	12	8	A
1-Naphthylamine	134-32-7	SPEC	1	1	1	1	D
1-Nitronaphthalene	86-57-7	FLAM	12	7	4	5	C
Sodium hydroxide	1310-73-2	ALK	7	2	6	3	E
Sulfuric acid	7664-93-9	OXID	8	5	3	4	E

[a] FLAM indicates the room for storage of flammable and combustible liquids, ALK for alkaline materials, OXID for oxidizers and mineral acids, and SPEC a room for storage of special hazards such as carcinogens.

tion for storage and retrieval of each container. As an example, the inventory list in Table 3.4 shows that 1-Nitronaphthalene is stored in the FLAMMABLE ROOM, Aisle 12, Rack 7, on Shelf 4 in Row 5 at Position C. To find a chemical or store it for easy retrieval, the first step is to locate the room, aisle, and rack. Within the rack the next step is to go to the assigned shelf or drawer, and then to go across to the row and back to the assigned position in the row. The last step is to store or retrieve the container.

An additional advantage of this system is that storage space can be provided for new chemicals without having to leave empty spaces in the original storage configuration or having to rearrange existing storage to make room for the new.

CHAPTER 4

INCOMPATIBLE CHEMICALS IN THE STOREROOM: IDENTIFICATION AND SEGREGATION

LESLIE BRETHERICK
Laboratory Safety and Design Consultant
Chemical Safety Matters
Dorset, England

4.1 INTRODUCTION

4.1.1 Previous Developments

The need to exercise control over the physical arrangement of chemicals in storerooms so as to minimize the consequences of accidental mixing of incompatibles by spillage or breakage, or in storeroom fire, has been recognized for a long time. One of the earliest published references to this need was that of Davison (1), who in 1925 proposed that safety considerations should be included when planning and implementing educational chemical storage facilities. In 1951 the Los Angeles Fire Department composed a Dangerous Chemicals Code, which still forms the basis of the list for segregating incompatible chemicals in the recent ACS booklet (2). An extended list of incompatible chemicals by Fawcett (3) has been reproduced extensively since it appeared in 1952.

Voegelein (4) gave a considerable amount of detail on the containers and storage conditions appropriate to a fairly wide range of dangerous chemicals, but only very general guidance on their segregation in storage. The most recently published guide, that by Pipitone and Hedberg (5), covers in detail many aspects of safe storage of chemicals on the relatively small scale. This reference again emphasizes the point made 18 years previously by Steere (6) that many potential problems may arise from arranging containers of chemicals in alphabetical order and ignoring incompatibilities.

The advent of new technology holds promise in providing additional data on the hazard potential of chemicals in storage. Using a computer program entitled CHETAH, Coffee (7) and Treweek (8) have developed mathematical procedures for predicting the self-explosive potential of organic chemicals and the energy release from binary systems of incompatible chemicals. A new and enhanced version 4.4 of CHETAH incorporating these improvements and capable of running on a personal computer is now available (9).

In spite of these reports and much more good advice that has long been available, the results of recent surveys (5,10) show that any reasonable level of segregation in storage of chemicals appears exceptional among many professional chemists and educators.

4.1.2 Need for a Positive Simple Basis

While pondering on this obviously undesirable state of affairs, two factors emerged as possible major contributors: (a) the preponderance of the almost universally negative aspects of advice available on chemical segregation—there is unlimited advice on what chemicals cannot be stored together, but infinitesimal information on those chemicals that can be stored together in relative safety; and (b) the considerable degree of complication and uncertainty that surrounds the question of how best to classify chemicals to allow a suitable system of segregation in storage to be developed.

There seems to be no clear consensus on what and how many classes or groups of chemicals exist that need to be segregated. Ten commonly mentioned incompatible groups are flammables, oxidants, reducers, concentrated acids and bases, water-reactives, toxics, peroxidizables, pyrophorics, and cylinder gases. The incompatibilities in the first five groups arise from the potential for exothermic, violent, or even explosive reactions on accidental intermixing. The presence of the sixth group of water-reactive compounds during water-based fire-fighting operations could lead to severe complications. Toxic materials often need physical control on their distribution for use, and for those of high volatility, special ventilation may be required. Peroxidizable materials need cool, dark, and (usually) air-exclusive storage, whereas pyrophorics effectively have a built-in ignition system, needing only contact with air (or sometimes water) to establish the triangle of fire. The tenth group requiring segregation, the cylinder gases, is exceptional in that as well as the hazard inherent in the contents of a particular cylinder, there is often a high kinetic energy content due to the state of compression of the contained gas.

However, these 10 groups are not mutually exclusive. Table 4.1 gives some examples of chemicals that belong simultaneously to at least two of these groups and that would, therefore, be difficult to classify for segregation solely on that basis. An additional assessment of the major hazard would be necessary.

More elaborate classification systems were proposed for classifying chemicals for segregation in storage, especially on the large scale. The J. T. Baker chemical safety course, based on an academic study (11), suggests 16 classes with considerable emphasis on inorganic–organic differentiation. The U.S. Coast Guard, considering the rather special requirements for safe shipboard and often bulk transportation and storage, uses no fewer than 43 classes (with

TABLE 4.1 Group Classification Problems of Chemicals[a]

Groups	Flammable	Acid	Base	Oxidant	Reducer	Gas	Toxic	Peroxidizable	Pyrophoric
Water-reactive	Acetyl chloride	Chlorosulfuric	Potassium hydroxide	Chromyl chloride	Lithium aluminum hydride	Boron trichloride	Phosgene	Acryloyl chloride	Trimethyl-aluminum
Flammable		Thio-acetic	Methyl-amine	Calcium hypochlorite	Calcium hydride	Butane	Hydrogen cyanide	Tetrahydro-furan	All[b]
Acid			Betaine	Nitric acid	Formic acid	Hydrogen chloride	Hydrofluoric acid	Acrylic acid	
Base					Hydrazine	Ammonia	Dimethyl-amine		Trisilyl-amine
Oxidant					Redox salts	Fluorine	Bromine		
Reducer						Hydrogen	Hydrazine		Titanium hydride
Gas							Arsine	Vinylidene chloride	Diborane
Toxic								Acrylo-nitrile	Phosphine
Peroxi-dizable									All[b]

[a]Difficulties in classifying hazardous chemicals using the usual suggested groups are listed in columns 2 through 9. Each of the chemicals shown belongs to at least two groups simultaneously.

[b]Note that all pyrophorics are flammable and peroxidizable, but the converse is not true.

89

several additional restrictions) as the basis of the CHRIS proposals for complete segregation (12).

Such a high degree of segregation is clearly impracticable and largely irrelevant to the needs of a relatively moderate-sized storage facility necessary to serve a small to medium group of laboratories engaged on relatively small-scale research or teaching activities.

4.2   MAJOR CRITERIA FOR SEGREGATION

4.2.1   Flammability

Of the total number of accidents involving the storage of chemicals, the most serious fraction involves fire as the initial acute hazard, usually causing considerable financial loss and occasionally injury or fatality. The possibility of fire and the implications of measures for effective prevention or control should therefore play a major part in the overall strategy of segregating chemicals in storage. This is not a new concept (13), but it does not seem to have been applied widely other than to stores (often bulk stores) for highly flammable organic solvents.

4.2.2   Water Compatibility

Water is the most appropriate and effective extinguishing medium for some types of fire involving chemicals, but the least appropriate to other types, so compatibility with water emerges as a further important criterion for storage considerations. Those combustible materials, liquid or solid, that are soluble in or denser than cold water are considered to be *water compatible*. Those combustible materials, liquid or solid, that are insoluble in and are less dense than (i.e., float on) cold water are considered *water incompatible*, as are those incombustible chemicals that react adversely with water. Special fire extinguishers (14) are necessary for some combustibles incompatible with water, such as metallic sodium or calcium hydride.

Although suppliers' catalogs or the labels on currently supplied containers of chemicals will usually indicate flammability, toxicity, and sometimes the density of the contents, it may be necessary to obtain water solubility and reactivity data from a literature source (15–21). Thus a decision on compatibility or incompatibility with water can be made properly with any further subclassification aspects necessary for the scheme outlined below.

4.3   IDENTIFICATION AND SEGREGATION OF HAZARDS

4.3.1   Identification and Classification

In view of its importance in overall storage considerations, flammability has been adopted as the major basis of division of the hazard classification proposed in this chapter. For reasons outlined previously, compatibility with water has been adopted as the next most important basis. Taken in combination, four major areas are produced:

Area 1   Flammables that are compatible with water
Area 2   Flammables that are incompatible with water
Area 3   Nonflammables that are compatible with water
Area 4   Nonflammables that are incompatible with water

Because the risk of fire in flammable materials is closely related to the volatility (as defined by the flash point), the criterion of flammability for the present purpose has been set, not on the absolute basis of combustibility or noncombustibility, but rather on a relative basis. Thus, those materials (usually organic liquids) with flash points below 37.8 °C (100 °F, Class IA, B, or C, labeled Flammables) in the United States, or below 55 °C (131 °F, labeled Extremely Flammable, Highly Flammable, or Flammable) in Europe, are included within the flammable grouping of Areas 1 and 2 for the present purpose. Materials with flash points above those values are included as combustibles of relatively low fire hazard, which may be stored with noncombustible materials of the same water-compatibility type in Areas 3 and 4.

Some degree of further physical segregation is required within Areas 1–4 for those toxic or reactive materials that need more closely controlled storage conditions than normal containers on open shelves. This aspect is discussed later.

There are four other groups that require separated and specific storage conditions. Three of these groups have major potential hazards, although limited quantities are usually kept in storage. These are:

Area 5   Materials that become unstable above ambient temperatures
Area 6   Materials unstable (or too volatile) at ambient temperature and that require refrigerated storage
Area 7   Pyrophoric materials for which loss of containment will cause ignition (see Section 4.1.2)
Area 8   Compressed gases in cylinders (special hazard; see Section 4.1.2)

Some degree of further chemical segregation usually will be necessary within Areas 5, 6, and 8. If a moderate amount of high-hazard material is necessary for the work in hand, or if many smaller quantities of different materials of similarly high hazard are to be stored, those chemicals may need storage in several separate cupboards or enclosures within the appropriate area.

4.3.2   Main Levels of Segregation

The eight main groups identified above are shown in schematic form as storage areas in Figure 4.1. With some secondary subdivision and separation,

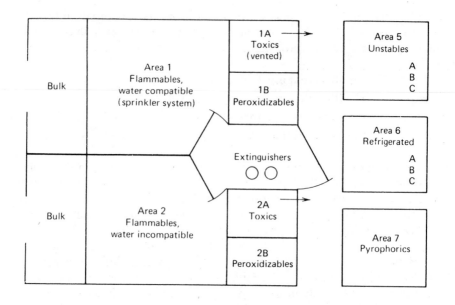

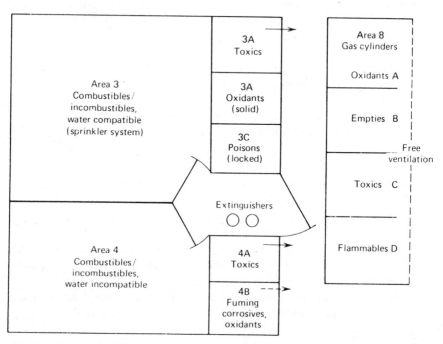

FIGURE 4.1 Outline of schematic arrangements for segregated storage of a wide range of laboratory chemicals. (*Note:* This is not intended as a working ground plan.)

the author proposes that those main groups form the basis of an effective segregation plan for a wide range of chemicals in up to kilogram quantities. These amounts are expected to be necessary for a group of laboratories engaged in normal small-scale chemical research or teaching activities. Specialized laboratories requiring larger quantities or proportions of particularly hazardous materials would need to adapt the general scheme shown in Figure 4.1 by accentuating the appropriate area, even to the extent of complete separation to accommodate the greater chemical volume.

It is emphasized that the schematic layout of Figure 4.1 is intended only to illustrate the general principles and is *not* a ground plan for a practical layout (see Section 4.4.2). Relatively cool temperate conditions, free from direct sun exposure and excessive cold in winter, are assumed to prevail in all areas except Area 6.

If the flameproofed Areas 1 and 2 are employed to accommodate bulk amounts of flammable solvents (i.e., in large quantities not easily handled as laboratory packages, as in the case of 55-gallon drums), storage and dispensing of those solvents should be separated from the main storage areas, minimally by fire-resistant partitions. Those bulk areas must be equipped with drum grounding facilities and accessed by means of separate wide doors for adequate drum handling. Such bulk areas are outlined schematically in Figure 4.1.

The relatively large size of Area 3 reflects the probability that the great majority of relatively nonhazardous solid (and particularly inorganic) materials would tend to be stored in a water-sprinklered area.

4.3.3 Sublevels of Segregation within a Storage Area

What is to be done with those chemicals that belong to the same area but are incompatible with each other? The author has adopted a practical point to simplify some of the segregation that should occur *within* each area. There is a relatively low probability that a solid, spilled or released through breakage, will effectively commingle with another incompatible solid because most solids do not flow readily. On the other hand, there is a higher probability that two incompatible liquids, spilled into accidental contact, will mix effectively with deleterious results.

Therefore, there could be overall advantage in storing solids separately from, and on shelves above, liquid chemicals within the same compatibility class. Furthermore, the priority to segregate pairs of incompatible liquids, and solid–liquid incompatible combinations, is higher than to segregate pairs of solids that may not be compatible.

4.3.3.1 *Secondary Areas 1–4*   The enclosed fire-resistant storage areas for toxic materials (Areas 1A, 2A, 3A, and 4A) shown vented to outside would need a modest ventilation rate. Dispensing of toxic materials from the storage containers is left to be done in well-ventilated laboratory hoods.

Deciding whether a material is of sufficient toxicity to merit removal from an open-shelf location and placement into the enclosed toxic storage area could be based on the vapor hazard index (VHI) recently proposed by Pitt (22). The index is a measure of the factor by which the concentration of a substance in a vapor-saturated atmosphere at a particular temperature exceeds the threshold limit value (TLV) or permissible exposure limit (PEL). The temperature in this application would be a typical ambient value, 20 or 25 °C. Use of the VHI rather than TLV or PEL as the decision basis for segregation would then include an indication of the risk of inhalation likely to be present in a spill or breakage incident in storage. The VHI can, of course, be calculated only for those several hundred common materials for which TLV or PEL figures are available.

Although the enclosed, cool and dark Areas 1B and 2B for storage of flammable peroxidizable compounds are schematically shown the same size as the adjacent toxic storage enclosures seen in Figure 4.1, the actual sizes are determined by the ratio of contents necessary for the particular laboratory or laboratories using the storage area. If the absolute quantities of toxic materials and peroxidizables were very small, a feasible arrangement would be to use an unvented dark lower part of a single cupboard for peroxidizables, with the toxics located in the separate, vented top section.

The separate locked cupboard representing Area 3C (for poisons) is proposed as a specific provision for those highly toxic or carcinogenic materials requiring mandatory control on access and use. In some cases where nonvolatile poisons are used, it may be more appropriate for Area 3C to be located in a laboratory under senior supervision, so that necessary registration formalities can be effected.

An interesting point emerged during consideration of the classification and segregation of oxidizers, which previously have tended to be lumped together as one internally compatible group. There is, in fact, need to exercise considerable care, and there seems a good case for segregating not only water-compatible oxidizers (Area 3B) from water-incompatible oxidizers (Area 4B) but also solid oxidizers from liquid oxidizers within both these groups on the grounds of potential reactivity.

Area 4B is devoted to the storage of materials (usually chemicals containing halogen and including some oxidizers) that react with moisture to produce toxic fumes that are also corrosive to structural materials. Ideally, the primary containers of such materials should not only be sealed with a reagentproof closure, but also be contained (not necessarily singly) within a secondary desiccated enclosure to minimize deterioration of the contents and concomitant fume generation. A proprietary "static hood" (23) is available for this purpose, but snap-lid plastic food containers with silica gel and some granular active carbon will serve equally well, particularly if the lid is also sealed round with PVC (not cellulose) adhesive tape. If such storage measures are used in Area 4B, the ventilation via negative pressure hinted at by the dotted arrow in Figure 4.1 should be unnecessary. Spent desiccant and carbon from the secondary storage boxes should be discarded rather than regenerated by heating. If total quantities are small, Areas 4A and 4B might be consolidated, using shelf segregation, into a single ventilated enclosure.

4.3.3.2 Secondary Areas 5–8   For materials stored in Area 5 (those unstable at elevated temperatures), the main requirement is isolation from heat sources. A location remote from fire-risk areas is essential. This is also true of Area 8. Adjacent placing of those two areas might be feasible.

The refrigerated Area 6, for storage of materials unstable at ambient temperature, or highly volatile toxics, will inevitably be a refrigerator or cold room, with two essential features: (a) a spark-free interior (external thermostat contacts) and (b) an individual power supply to minimize the risk of power failure and subsequent warming. Refrigerated storage is often provided in laboratories for convenience in supervision, but if located remotely, some device to warn personnel of power failure is necessary so that remedial action can be taken. Minimal internal segregation of materials in the refrigerated store would require separate shelf trays (with edge spaces for cold circulation), perhaps with additional secondary containment for volatile corrosives as in Area 4B. Additional refrigerators will be necessary for storage of more than minimal quantities of highly incompatible thermally unstable materials.

Since Area 7 (for storing pyrophoric materials) has the potential for high fire risk, separation by a safe distance from flammable or heat-sensitive materials is of paramount importance. A considerable degree of secondary containment, preferably under dry inert atmosphere (nitrogen-filled desiccators), will be necessary to prolong the storage life of most highly reactive materials of this group. Subsidiary storage of pyrophorics in laboratories is also likely to occur, because small (working) quantities of pyrophoric reagents and catalysts are usually maintained for safety and convenience in use in the inerted glove boxes used in laboratories engaged in organometallic chemistry. It seems essential to exclude from Area 7 those few pyrophoric materials (white phosphorus, zirconium powder) that need water cover for safe storage, in view of the water incompatibility of most pyrophorics. No dispensing of pyrophorics is permissible in storage Area 7.

Isolation from potential fire sources is the main requirement for cylinder storage in Area 8, to minimize the possibility of cylinder rupture and release of the often highly compressed gaseous contents. A secondary requirement is that the whole area must be freely ventilated to permit the dispersion of any toxic or flammable gas that might leak slowly through a faulty cylinder valve, which is not an uncommon occurrence in practice. Segregation of cylinders into the four indicated sections is desirable if more than a dozen or so cylinders are stored. The specific area for "empties" (which should invariably contain a small residual pressure to minimize the possibility of air contamination of the cylinder) is an important feature of the arrangement. The securing of cylinders by retaining straps or chains is equally important in storage as it is in the use of cylinders. Cylinders of inert gases may be dispersed as convenient into the other full cylinder storage sections.

4.4 PRACTICAL APPLICATIONS

4.4.1 Classification Applied to a Range of Chemicals

To check the general validity of the proposal developed above and to provide examples, particularly of the sublevels of segregation that may be necessary within each area, a selection of some 200 chemicals that might be expected to be found in typical storerooms serving multiproject research laboratories has been classified into the eight major areas discussed. Main levels of segregation within each area, and any points of detail related to these, are indicated in the groupings below by the following abbreviations (note that each prefix also applies to all unprefixed items listed below it):

o = Oxidizer	b = Base			p = Peroxidizable
r = Reducer	n = Not otherwise classified			d = Store dark
a = Acid				

Specific incompatibilities are indicated by an abbreviation in parentheses following the name of the chemical:

AREA 1 Flammables (Compatible with Water)

Solids	Liquids	
o Ammonium dichromate	o Nitromethane (no b)	Acetonitrile
Sulfur powder	r Hydrazine hydrate (b)	Chlorobenzene
	a Acetic acid	Methanol
	Phosphinic acid (r)	Methyl isobutyl
	b *tert*-Butylamine	ketone
	Pyridine	Propanol
	n Acetone	

1A—Toxics

SOLID

Acrylamide (d)

LIQUIDS

Acetone cyanohydrin (no a,b)
Acrylaladehyde (p)
Allyl alcohol
Allylamine
Allyl chloride
Allyl chloroformate
Carbon disulfide
Epichlorohydrin

1B—Peroxidizables

Crotonaldehyde
1,1-Dimethoxyethane
1,2-Dimethoxyethane
Dioxane
Ethyl acrylate
Tetrahydrofuran

AREA 2   Flammables (Incompatible with Water)

Solids	Liquids
r Aluminum powder	Acetic anhydride
Calcium hydride	Acetylacetone
Lithium aluminum hydride	Amyl acetate
Magnesium powder or turning	*tert*-Butyl chloride
Sodium hydride	Cyclohexane
Sodium dithionite	Ethyl acetate
n Phosphorus pentasulfide	Petroleum ether
	Toluene

2A—Toxics

2B—Peroxidizables

	SOLID	LIQUIDS
Acetyl chloride (store dry)	Potassium (no CO$_2$)	Dibutyl ether
Acrylonitrile (p)		Diethyl ether
Benzene		Styrene
Toluene-2,4-diisocyanate		Vinyl acetate
		Vinylidene chloride

AREA 3   Combustibles and Incombustibles (Compatible with Water)

Solids	Liquids
r Hydrazine sulfate	o Sodium hypochlorite (no a)
Hydroxylamine hydrochloride	Nitric acid (a)
a Adipic acid	Perchloric acid (a)
Benzoic acid	r Formaldehyde solution (no HCl)
Chloroacetic acid	Formic acid (a)
Citric acid	a Hydrochloric acid (no o)
Cyanoacetic acid	Phosphoric acid
Maleic anhydride (no b)	b Ammonia solutions
Oxalic acid	Butylamine
Phenol	Cyclohexylamine
Sulfamic acid	Ethanolamine
p-Toluenesulfonic acid	Morpholine
b 2-Aminopyridine	b Sodium hydroxide solutions
Calcium oxide	n Acetophenone
n Aluminum oxide	Aniline
Aluminum sulfate	Butoxyethanol
o-Aminophenol	Carbon tetrachloride
Ammonium chloride	Chloroform
Ammonium fluoride	Dibutyl phthalate
Ammonium sulfate	Furfural
Ammonium thiocyanate (d)	Nitrobenzene
Anthracene	1,1,1-Trichloroethylene

(Continued)

Solids	Liquids
Biphenyl	
2,2-Bis(4-hydroxyphenyl)propane	
Calcium carbonate	
Caprolactam	
p-Chlorophenol	
Cobalt(II) acetate	
Copper sulfate	
p-Nitrotoluene	
Sodium azide (no a)	
Sodium sulfide (no a)	
Zinc powder	
Zinc chloride	

3A—Toxics

SOLIDS	LIQUIDS
2,4-Dinitrochlorobenzene	Acrylic acid (p)
	Adiponitrile
	Benzenethiol (stench)
	Benzonitrile
	Benzoyl chloride
	Bromine
	Chloropicrin
	Dimethylformamide
	Dimethyl sulfate
	Hexamethylphosphoramide

3B—Solid Oxidizers (Liquids in Area 3)

Ammonium nitrate	Manganese dioxide
Ammonium perchlorate	Potassium chlorate
Ammonium persulfate	Potassium dichromate
Barium peroxide	Potassium nitrate
Calcium hypochlorite	Potassium permanganate
Chromium trioxide	Silver nitrate (d)
Dichloroisocyanuric acid	Sodium nitrate
Lead nitrate	Uranyl nitrate
Magnesium perchlorate	

3C—Poisons (Including Carcinogens, in Locked Cupboard)

Arsenic trioxide	Cadmium oxide
Barium chloride	β-Naphthylamine
Benzidine	Potassium cyanide
Beryllium oxide	Thallium nitrate

AREA 4 Combustibles and Incombustibles (Incompatible with Water)

Solids

o Sodium peroxide	o Sulfuric acid	Cyclohexane
r Calcium	n Anisole	Decahydronaphthalene
Sodium borohydride	Benzaldehyde	Dimethylaniline
b Potassium hydroxide	Benzyl alcohol	Toluidine
	Butanol	Xylidine
	Cumene	

4A—Toxics

SOLID	LIQUID
γ-Hexachlorocyclohexane	Dimethylcarbamoyl chloride

4B—Fuming Corrosives (Store Dry)

SOLIDS	LIQUIDS	
Aluminum chloride	Acetyl bromide	Methanesulfonyl chloride
Antimony trichloride	Antimony pentachloride	Oleum
Cyanogen bromide	Benzotrichloride	Phosphorus tribomide
p-Toluenesulfonyl	Benzoyl chloride	Phosphoryl chloride
chloride	Chlorosulfuric acid	Silicon tetrachloride
	Chromyl chloride (o)	Sulfinyl chloride
	Ethyltrichlorosilane	Titanium tetrachloride

AREA 5 Materials Unstable above Ambient Temperature

5A—Oxidizers	*5B—Organics*
Cumene hydroperoxide	m-Dinitrobenzene
Hydrogen peroxide (vent container)	Picric acid (store wet)
Peroxyacetic acid	

AREA 6 Refrigerated Materials Unstable or Too Volatile at Ambient Temperature

6A—Oxidizers	*6B—Toxics*	*6C—Flammables*
Dibenzoyl peroxide	Methyl fluorosulfate	Acetaldehyde
Di-*tert*-butyl peroxydicarbonate	Methyl iodide	Dimethyl ether

AREA 7 Pyrophorics

Butyllithium solutions
Diethylzinc
Triethylaluminum solutions

AREA 8 Cylinders of Compressed or Liquefied Gases

8A—Oxidizers	8B—"Empties" (Slight residual pressure)	8C—Toxics	8D—Flammables
Chlorine		Arsine	Acetylene
Oxygen		Boron trichloride	Ethylene oxide
Ozone solutions		Phosphine	Hydrogen
		Sulfur dioxide	

4.4.2 Further Practical Necessities

As has been stressed previously, the general arrangement presented in Figure 4.1 and discussed in some detail above deals with the development of a general principle for the segregation of chemicals in small-scale storage.

To apply this principle practically to an existing storage facility or to design a new one will, especially in the latter case, require much detailed consultation and consideration of many specific factors including local geography, codes, and fire regulations. Some of the major practical factors are as follows, with pertinent references given in parentheses: planning and design of chemical storage facilities (24, 25), segregation of chemicals (2, 3, 17–20, 26), fire protection in chemical storage (27, 28), chemical container and labeling requirements (17, 29), record keeping (5), and storage of gas cylinders (30, 31).

REFERENCES

1. H. F. Davison (1925) *J. Chem. Educ.* **2**, 782.
2. ACS Committee on Chemical Safety (1985). *Safety in Academic Chemical Laboratories,* 4th ed., ACS, Washington, DC, Appendix 3.
3. H. H. Fawcett (1952). *Chem. Eng. News* **30**, 2588.
4. J. F. Voegelein (1966). *J. Chem. Educ.* **43**, A151.
5. D. A. Pipitone and D. D. Hedberg, (1982). *J. Chem. Educ.* **59**, A159.
6. N. V. Steere (1964). *J. Chem. Educ.* **41**, A859.
7. R. D. Coffee (1972). *J. Chem. Educ.* **49**(6), A343.
8. D. N. Treweek (1980). *Ohio J. Sci.* **80**(4), 160.
9. D. J. Frurip, E. Freedman and G. R. Hertel (1989). *Plant/Oper. Prog.* **8**(2), 100.

10. R. Powers and P. Redden (1982). *J. Chem. Educ.* **59**, A9.

11. S. H. Pouliot (1973). *A Program for Compatible Chemical Storage of Chemicals,* Thesis, University of North Carolina.

12. *Chemical Hazard Response Information System (CHRIS) Manual,* 3d ed., U.S. Coastguard, Washington, DC, 1985.

13. Reference 5, p. A161.

14. *Hazardous Chemicals Data, NFPA 49,* 3d ed., National Fire Protection Association, Boston, MA, 1975.

15. R. C. Weast and M. J. Astle (Eds.), *Handbook of Chemistry and Physics,* (new editions annually), CRC Press, Boca Raton, FL.

16. M. Windholz (Ed.) (1989). *The Merck Index,* 11th ed., Merck Company, Rahway, NJ.

17. L. Bretherick (Ed.) (1986). *Hazards in the Chemical Laboratory,* 4th ed., Royal Society of Chemistry, London.

18. *Handling Chemicals Safely, 1980,* 2d (English) ed., Dutch Association of Safety Experts/Dutch Chemical Industry Association/Dutch Safety Institute, Amsterdam, 1980.

19. *Manual of Hazardous Chemical Reactions NFPA 491M;* 5th ed., National Fire Protection, Boston, MA, 1975. Reprinted in larger format, 1985.

20. L. Bretherick (1990). *Handbook of Reactive Chemical Hazards,* 4th ed., Butterworths, Boston, MA.

21. N. I. Sax and R. J. Lewis (1988). *Dangerous Properties of Industrial Materials,* 7th ed., Van Nostrand-Reinhold, New York.

22. M. J. Pitt (1982). *Chem. Ind. (Lond.),* 804.

23. Static Hood from Bel Art Products, Pequannock, NJ 07440. Construction details for a storage cabinet have been published: Di Berardinis, L. et al. (1983). *Am. Ind. Hyg. Assoc. J.* **44**(8), 585–588.

24. K. Everett and D. Hughes (1975). *Guide to Laboratory Design,* Butterworths, London, pp. 104–111.

25. D. W. Shive and W. H. Norton (1980). *Hazardous Chemical Safety* (course notes), J. T. Baker Chemical Co., Phillipsburg, NJ, pp. 16.1–16.8.

26. *Prudent Practices for Handling Hazardous Chemicals in Laboratories,* National Research Council, Washington, DC, 1981, pp. 215–229.

27. N. V. Steere (1971). *Handbook of Laboratory Safety,* 2d ed., CRC Press, Cleveland, OH, pp. 179–199.

28. *NFPA Fire Codes,* 43A, B, C; 45, National Fire Protection Association, Quincy, MA 1986.

29. M. E. Green and A. Turk (1978). *Safety in Working with Chemicals,* MacMillan, New York, pp. 16–17, 35–39, 117–119.

30. Reference 26, pp. 574–578.

31. W. Braker and A. L. Mossman (1981). *Matheson Gas Data Book,* 6th ed., Matheson Gas Products, East Rutherford, NJ.

CHAPTER 5

LABELING PRACTICES FOR CHEMICALS IN LABORATORIES, WORKPLACES, AND CHEMICAL STOREROOMS

L. JEWEL NICHOLLS
Environmental and Safety Chemist
University of Illinois at Chicago
Chicago, Illinois

5.1 INTRODUCTION

Proper labeling and adequate storage of containers of chemicals are vital to the certainty with which we can use those chemicals. Labels attached to any container of chemicals at the factory assure us of the quality, grade, and limits to impurities *as it is being packaged*.

The label in most cases holds the key to the quality of the contents for those who use them, as well as other valuable and essential information for those who only handle the containers. Therefore, the original label must be kept in good condition, and some information may need to be added so that continuing quality may be assessed properly. The label can also be a key to proper storage for (a) easy retrieval and (b) optimum shelf life. Such information is important in the event of a spill for proper assessment of hazards to prevent injury and facilitate proper cleanup. Eventually the information will help with decisions about disposal.

5.1.1 Historical Background

The Manufacturing Chemists Association (MCA, now CMA), the National Fire Protection Association (NFPA), the Bureau of Standards, and various companies have developed systems to improve labels on hazardous chemicals, using colors and number codes. In 1966 the NFPA (1) published a four-diamond system using four boxes in a diamond shape. They indicate health hazard (left, blue), fire hazard (top, red), instability hazard (right, yellow), and

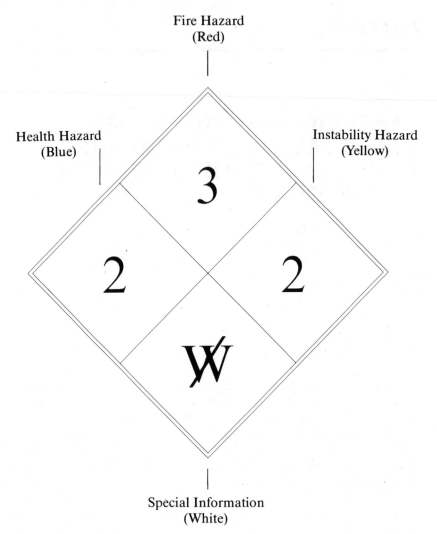

FIGURE 5.1   NFPA diamond hazard label.

special information (bottom, white) (Figure 5.1). A number code for severity of hazard was part of this system using a scale of 0 to 4, with 4 being the most severe and 0 meaning no hazard at all, even in a fire.

After the OSHA was enacted in 1970, it became clear that the language was vague about the ways in which hazards were to be communicated to workers. As a result, a section of the OSH Act was added, 29CFR Part 1910.1200, called the Hazard Communication Standard (2), in which the methods of communication of such hazards were spelled out. These included Material Safety Data Sheets (MSDSs), labels, and employee training. The Hazard Communication Standard was adopted in 1983 and applied to manufacturing industries in which there were the most injuries (48 FR 53280) (3). By May 1988, it was extended to all workplaces in the United States, with exceptions for most laboratories (52 FR 31852). The Hazard Communication Standard mandated that labels must contain not only the identity of the chemical but information about hazards and precautions as well. Various companies, following the lead of J. T. Baker Inc., have developed a system of pictograms for better communication of chemical hazards, especially for workers with poor understanding of written English, and for emergency response (see the following).

5.1.2   Today's Improved Labels

Labels today contain much more information than they did three decades ago. Labels contain chemical name, hazard warnings, and emergency information as well as information about quality and limits of impurities, suggested protective equipment, expiration dates, physical data such as melting or boiling point, chemical abstracts number, and spectroscopic data.

On large containers hazards are both listed and depicted by pictograms on the labels. The Baker SAF-T-DATA system of pictograms is based on the National Bureau of Standards' list of hazards. There are symbols and severity number codes for many different hazards. These are illustrated and explained in Figure 5.2. Laboratory protective equipment pictograms for labels are illustrated in Figure 5.3. A complete J. T. Baker Inc. label for nitric acid is shown in Figure 5.4. Health, flammability, reactivity, and contact hazards are indicated by symbol, number indicating severity, and the word indicating severity. Protective clothing is indicated by symbols as well as words.

Fisher Scientific Company uses a system called ChemAlert. It uses very basic symbols and an NFPA hazard diamond. The symbols involve (a) the Fisher Man and the variations of him with the personal protection needed and (b) the symbol of a fumehood if ventilation is required. Words are also used to show the suggested protective equipment (Figures 5.5 and 5.6). Most workers are protected by right-to-know laws which require this. Besides, chemical companies find that customer service is very important in an attempt to help people to use and store the material safely and efficiently.

5.2   REGULATORY NEED FOR LABELING

The OSHA was first passed in 1970 and is Title 29 of the Code of Federal Regulations (CFR) Part 1910 (2). It included authority for the Secretary of Labor to "prescribe the use of labels or other appropriate forms of warning as are necessary to insure that employees are apprised of all hazards." In 1983 (48 FR 53280) these rules became Section 1200 of that title, 29 CFR 1910.1200, also known as the Hazard Communication Standard, and applied to only 20 industrial classifications (SIC codes); but by 1987 (52 FR 31852) the coverage

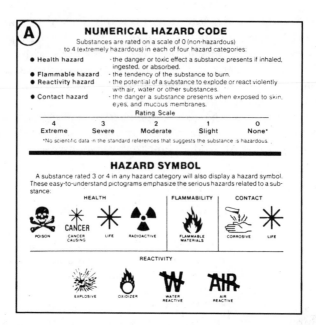

FIGURE 5.2 Baker SAF-T-DATA guide to hazards. (Used with permission of J.T. Baker Inc., Phillipsburg, NJ.)

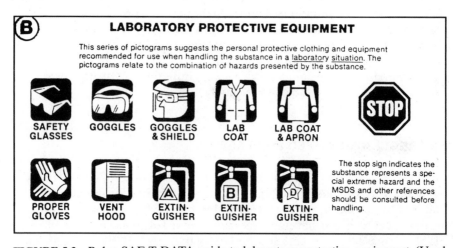

FIGURE 5.3 Baker SAF-T-DATA guide to laboratory protective equipment. (Used with permission of J.T. Baker Inc., Phillipsburg, NJ.)

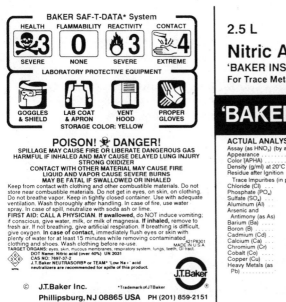

BAKER SAF-T-DATA* System

HEALTH	FLAMMABILITY	REACTIVITY	CONTACT
3	0	3	4
SEVERE	NONE	SEVERE	EXTREME

LABORATORY PROTECTIVE EQUIPMENT

GOGGLES & SHIELD	LAB COAT & APRON	VENT HOOD	PROPER GLOVES

STORAGE COLOR: YELLOW

POISON! ☠ DANGER!
SPILLAGE MAY CAUSE FIRE OR LIBERATE DANGEROUS GAS
HARMFUL IF INHALED AND MAY CAUSE DELAYED LUNG INJURY
STRONG OXIDIZER
CONTACT WITH OTHER MATERIAL MAY CAUSE FIRE
LIQUID AND VAPOR CAUSE SEVERE BURNS
MAY BE FATAL IF SWALLOWED OR INHALED

Keep from contact with clothing and other combustible materials. Do not store near combustible materials. Do not get in eyes, on skin, on clothing. Do not breathe vapor. Keep in tightly closed container. Use with adequate ventilation. Wash thoroughly after handling. In case of fire, use water spray. In case of spill, neutralize with soda ash or lime.
FIRST AID: CALL A PHYSICIAN. **If swallowed,** do NOT induce vomiting; if conscious, give water, milk, or milk of magnesia. **If inhaled,** remove to fresh air. If not breathing, give artificial respiration. If breathing is difficult, give oxygen. **In case of contact,** immediately flush eyes or skin with plenty of water for at least 15 minutes while removing contaminated clothing and shoes. Wash clothing before re-use.
TARGET ORGANS: eyes, skin, mucous membranes, respiratory system, lungs, teeth, GI tract.
DOT Class: Nitric acid (over 40%) UN 2031
CAS NO: 7697-37-2
J.T.Baker NEUTRASORB® or TEAM* 'Low Na·' acid neutralizers are recommended for spills of this product.

© **J.T.Baker Inc.** *Trademark of J.T.Baker
Phillipsburg, NJ 08865 USA PH (201) 859-2151

2.5 L   9598-33

Nitric Acid, 70.0-71.0%
'BAKER INSTRA-ANALYZED'® Reagent
For Trace Metal Analysis

HNO₃ FW 63.01

'BAKER ANALYZED'®

ACTUAL ANALYSIS, LOT C07039

Assay (as HNO₃) (by acid-base titrn)	70.9	%
Appearance	Passes Test	
Color (APHA)	< 5	
Density (g/ml) at 20°C	1.429	
Residue after Ignition	< 1	ppm

Trace Impurities (in ppm):

Chloride (Cl)	< 0.08	Iron (Fe)	0.01
Phosphate (PO₄)	< 0.1	Lead (Pb)	0.0004
Sulfate (SO₄)	< 0.5	Lithium (Li)	< 0.02
Aluminum (Al)	0.05	Magnesium (Mg)	0.001
Arsenic and		Manganese	
Antimony (as As)	< 0.005	(Mn) (by AAS)	0.0002
Barium (Ba)	< 0.02	Mercury (Hg)	< 0.005
Boron (B)	0.001	Nickel (Ni)	0.0002
Cadmium (Cd)	< 0.002	Potassium (K)	< 0.1
Calcium (Ca)	0.02	Silicon (Si)	0.006
Chromium (Cr)	< 0.0006	Silver (Ag)	< 0.0001
Cobalt (Co)	< 0.0002	Sodium (Na)	< 0.02
Copper (Cu)	0.0002	Strontium (Sr)	< 0.002
Heavy Metals (as		Tin (Sn)	< 0.001
Pb)	< 0.1	Zinc (Zn)	< 0.002

FIGURE 5.4 An example of a label on a J.T. Baker Inc. nitric acid bottle, showing hazard symbols and guide to protective equipment. (Used with permission of J.T. Baker Inc., Phillipsburg, NJ.)

Safety Codes

GOGGLES
GLOVES
APRON

HOOD

NFPA Hazard Codes*

FIGURE 5.5 Fisher Scientific Company's ChemAlert symbols for labels showing hazard codes and safety codes. Basic safety information is conveyed to the worker by means of the graphics depicting eye guard, proper gloves, safety clothing, and use of a fume hood. (Used with permission of Fisher Scientific Company, Pittsburgh, PA.)

Methanol
Methyl Alcohol; Wood Alcohol
CH_3OH

F.W. 32.04	Flash point 50°F
CAS-67-56-1	bp 64°C

ChemAlert™ Storage Code: Red. Danger! Flammable! Poison! Harmful if inhaled. May be fatal or cause blindness if swallowed. Cannot be made nonpoisonous. Narcotic. Irritant. Store in standard flammables cabinet.

Safety Precautions: Use fume hood. Wear eye guard, proper gloves, and safety clothing. Wash thoroughly after handling. Keep container closed. Keep away from heat, sparks, and open flames.

D.O.T. Flammable liquid.

Methyl Cellulose

CAS-9004-67-5

ChemAlert™ Storage Code: Gray. Caution! Harmful if swallowed.

Safety Precautions: Wear eye guard. Do not take internally.

FIGURE 5.6 Examples of how symbols from Fisher Scientific Company's ChemAlert system may be used on chemicals with toxic (or flammable) hazards and relatively nonhazardous chemicals. Examples of information that may be on labels are shown here for moderately hazardous and nonhazardous materials. The diamond to the right refers to the NFPA hazard codes. (Used with permission of Fisher Scientific Company, Pittsburgh, PA.)

was expanded to apply to all industries, including schools and service industries (3). Some states have their own communication standards which may have additional features. Other countries have developed similar systems.

The provisions are based on communication, both written and verbal. Training, in addition to labeling and Material Safety Data Sheets (MSDSs), is an important part of communication. Workers have to be told of their rights to ask for good labels and other information, and are to be held responsible, once trained, for proper safety techniques, much of it based on information on the labels.

Until 1990 laboratories were exempt from these regulations because of their diverse nature and the usually high level of professional training of the laboratory workers. Proposed rules for a performance-oriented laboratory standard were published in July 1986. After long delay and many comments, the final rule was published on January 31, 1990, as 29 CFR 1910.1450, part of Subpart Z. Its title is "Occupational Exposure to Toxic Substances in Laboratories," and it is otherwise known as the OSHA laboratory standard (55 FR 3300). It became effective May 1, 1990, and everyone must have been in compliance by January 1991. However, it is a performance-oriented standard. The storage requirements for laboratories include management of MSDSs received, environmental monitoring, and a system of housekeeping, maintenance, and inspections for proper management of the storage area. Each laboratory should develop a method for recordkeeping, labels, and MSDS storage to meet the needs of the laboratory.

The laboratory standard applies to the management of individual containers in laboratories by the laboratory workers, whereas the stockroom or storeroom down the hall, which is managed by a nonscientist, is under the regular OSHA rules. Any chemical container must have a proper label with as much information as possible for identification and hazard evaluation. The guidelines in *Prudent Practices for Handling Hazardous Chemicals in Laboratories* (4) are cited by page numbers.

5.3   LABORATORY NEED FOR LABELING

5.3.1   Purchased Chemicals

In laboratories, chemicals are purchased for specific purposes. In many cases the quality is important, and special grades are available for particular laboratory procedures, such as HPLC. However, no matter how well something is packaged and labeled, if impurities enter the container or if heat or electromagnetic energy are allowed to affect the contents, then the contents may undergo substantial change in a short time and be unsuitable for the purpose for which they were purchased.

5.3.2   Chemicals Synthesized in Laboratories

Chemicals synthesized in laboratories very often become a problem when they are not properly labeled. A page number in a laboratory notebook is inadequate. A probable structure is helpful. Melting or boiling points, stability information, reactivity, and IR peaks are all useful data that can assist in a laboratory clean-out or in future usage of these unique substances.

5.3.3   Continued Purity Assurance

Methods of characterization and purification are well documented (5). Many manufacturers of fine chemicals list references to laboratory methods such as IR, NMR, or FT–IR spectra in their catalogs and on labels of many of the fine chemicals they sell. Figure 5.4 shows a label with analysis of nitric acid. The Department of Transportation (DOT) code and Chemical Abstracts (CAS) number are listed. In some labels the Beilstein number and physical data such as boiling point are included. These references may be used at any time to compare quality after the container of the chemical has been on the shelf and reopened many times.

5.3.4 Protection of Older Stocks of Chemicals

Serious problems in laboratories occur because older bottles of reagents are identified only with a name, or because someone has protected a label with cellophane tape that has curled up, removing valuable information. In these cases chemical detective work must be done to establish purity. Additions to the label as to the quality of the reagent at a particular time are very useful data. Protection of the original label and auxiliary labels are important as well. Coats of lacquer and sheets of special adhesive label protector are especially designed for these purposes.

5.3.5 Special Hazards on Storage

Reagents known to decompose over time to dangerous states or by-products should have auxiliary labels added, with dates received and opened, and prominent warnings should be added to these labels. Modern labeling often indicates expiration dates for diethyl ether and other common peroxide formers. This practice is based on normal storage and usage, but it is not infallible. Inhibitors are added to many grades of ethers to retard reaction with oxygen. However, all containers of solvents that are peroxide formers could contain quantities of peroxides that, if concentrated by evaporation of the solvent, could explode with shock or the heat of evaporating. The label shown in Figure 5.7 may be prepared and added to such things as ethers or other peroxide formers. See Table 5.1 for a partial list of compounds that can form peroxides or react dangerously if exposed to peroxides and their approximate shelf lives (6).

DANGER: PEROXIDE-FORMER

Date received _____

Date opened _____

Use or dispose of by _____

Test before heating. Store away from heat and light.

FIGURE 5.7 Warning label for peroxide-forming chemicals.

TABLE 5.1 Solvents and Other Compounds Which Form
Peroxides or Are Affected by Peroxide Initiation

Compounds in Which Dangerous Amounts of Peroxides Can Form on Storage	
Isopropyl ether	Potassium metal
Vinylidine chloride	Sodium amide

Compounds Whose Peroxide Hazard Is Apparent on Concentration, as Evaporation or in a Distillation	
Diethyl ether	Vinyl ether
Dicyclohexane	Decalin
para-Dioxane	Diacetylene
Tetrahydrofuran	Tetralin
Methyl acetylene	Glyme
Cyclohexene	Organometallics
Vinyl ethers	Benzyl ethers
Acetal	Sodium and potassium alkoxides

Monomers Which Can Undergo Hazardous Polymerization Due to Peroxide Initialization	
Butadiene	Chloroprene
Styrene	Tetrafluoroethylene
Vinyl acetate	Vinyl acetylene
Vinyl chloride	Vinyl pyridine

Source: Jackson et al. (6).

5.4 SUPPORT SYSTEMS FOR LABORATORIES

5.4.1 Computer Systems

Computer systems are available for generating labels on pressure-sensitive stock. There are colorful stickers that may be attached to draw attention to hazards. The computer system may be tied to a database that, in addition to the name, formula, warnings, and first aid information, may be able to evoke such information as CAS number, stock number, spill cleanup directions, and storage code.

5.4.2 Repackaging Chemicals

Often reagents are dissolved or diluted to make stock solutions. Labels for dilutions and mixtures need to be prepared with care. They should contain names and concentrations of components, hazards, use of the mixture, and expiration date indicating the time when the mixture or solution will no longer be needed and should be disposed of. Any such label should indicate the name of the person making the solution.

5.4.3   Labels and Time-Related Degradation

A disappointment awaits the naive chemist who picks a bottle of organic liquid off the shelf only to find a black lumpy gel. Thermodynamically unstable compounds will eventually decompose to form more stable substances. An example is a reaction of aldehydes and ketones catalyzed by moisture to form the polymeric mixture noted above. Thus, there may be a mixture of polymer and impurities as well as the original aldehyde or ketone in the bottle. Fractional distillation may recover some of the original compound if the degradation has not gone too far (5).

5.4.4   Labels to Warn of Catalytic Effect of Light and Impurities

Reagents known to be light sensitive are now almost always labeled as such by the manufacturer. They are placed in metal or dark glass containers. Repackaging them in clear glass or storing them near a window is clearly dangerous. The degradation of a compound starts with free radical formation, sometimes with reassociation, but often with rearrangement of the free radical fragments to give impurities or to form polymers. Therefore, storage in dark containers away from windows and sources of light, especially ultraviolet, is essential.

5.5   SPECIAL CASES FOR LABELING

5.5.1   Labels for Sensitive Materials Needing Refrigeration

Organic peroxides are a class of compounds characterized by the $-O-O$ or oxygen–oxygen linkage. Many are too explosive to be prepared commercially (7). This is because of the low activation energy needed to form free radicals which then initiate self-decomposition or initiate decomposition or polymerization of other materials. Organic peroxides must be refrigerated and protected from heat, light, and sparks. The decomposition may be slow or uncontrolled. A sample of benzoyl peroxide found at the back of a refrigerator may have been prevented from accidental explosion, but its purity should be suspect.

Therefore, anything that has a short shelf life or that needs special storage conditions should be clearly labeled as such. A label such as that in Figure 5.8 should be added to the manufacturer's label. The dates received and opened as well as an expiration date (sometimes supplied by the manufacturer) should be clearly shown. Amounts over 25 grams should be stored in special outdoor storage rooms.

The *Aldrich Catalog* indicates that benzoyl peroxide cannot be shipped by Parcel Post or United Parcel Service (UPS) in the 50- or 500-gram sizes (8, p. 156). Larger sizes are not available. The 70% solution in water is offered, of which the 100-gram size may be shipped Parcel Post or UPS. Comparison of the thermodynamics of peroxides and other organic compounds indicates

---

MUST BE KEPT COLD

Store at 40°F or _____

Received _____

Use or dispose of by _____

Date opened _____

KEEP AWAY FROM SPARKS AND HEAT

FIGURE 5.8   Warning label for heat-sensitive chemicals.

that the energy of activation of explosion is substantially less for peroxides and thus less stable to tiny inputs of energy.

5.5.2   Labels for Reactive Chemicals

Any chemicals that have potential for explosive reaction should be stocked and stored as explosives (4, p. 62). Labels should denote the explosive characteristic. Picric acid has been used so widely in the past that it has been found in the cabinets and on the shelves of research laboratories, clinics, and high schools. People have reacted in panic, calling in bomb squads for very expensive disposal. Maintenance of at least 10% moisture in any explosive may keep it safe for use. Under arid conditions of storage, as in Arizona, the danger is far greater than in a Great Lakes state.

For compounds such as picric acid that become explosive when dry, it is a good idea to record the weight of the new bottle on the label before opening, and again each time before and after removing any material for use in the laboratory. Water should be added as a matter of course. If the tare weight has decreased markedly between usage, it may be attributed to loss of water and resulting instability. The label should clearly suggest the appropriate action to take so that the reagent can have an increased shelf life (Figure 5.9).

Several structural and thermodynamic properties may contribute to the explosive decomposition of many substances. Functional group nomenclature provides clues that may alert a chemist to instability. Chemical names that contain the prefixes, per-, peroxy-, azo-, diazo-, and acetyl- can alert one to the fragile bonds of explosives and other unstable molecules. A large amount of oxygen in an organic molecule may mean a large amount of energy released on decomposition along with gaseous products such as carbon monoxide, car-

---

EXPLOSIVE WHEN DRY

Weigh before opening and compare weight with previous tare.

Date	Mass before opening	Mass after opening	Initials
——————	————————————	————————————	——————
——————	————————————	————————————	——————
——————	————————————	————————————	——————
——————	————————————	————————————	——————

FIGURE 5.9 Warning label for reactive (explosive) chemicals.

bon dioxide, water, and possibly oxygen in its elemental or molecular form. Examples of such compounds are nitromethane, nitrosophenol, and inorganic salts such as perchlorates, chlorates, or nitrates of many metal ions.

5.5.3 Recycled Solvents and Reagents

The current cost of chemicals and, more importantly, the very high cost of disposal of them, makes it imperative that recycling and recovery techniques be used. Proper labels for recycled solvents are important. The date of recovery, the estimate or assurance of purity, and the name of the person who performed the recovery will help workers to reuse the material effectively. Additionally, purification or recovery of air- and light-sensitive chemicals will mean loss of any inhibitors. Redistilled peroxide formers must be used immediately. Testing for peroxides and removal of any peroxides must be completed before purification (6). It is especially important that those chemicals be packaged properly and *used immediately*. A label with pertinent information should be added. (An example is shown in Figure 5.10.)

5.5.4 Carcinogens

Carcinogens vary in their toxicity, but it is important that contact with them be limited and storage of them be secure so that unauthorized persons will not

RECYCLED SOLVENT

Identify _____

Quality _____

Precautions _____

_____

Date _____ Signature _____

FIGURE 5.10 Identification label for recycled solvents.

use them. Each container should be labeled CAUTION: HIGH CHRONIC TOXICITY OR CANCER SUSPECT AGENT. The storage area for them should also be labeled with a warning and should have limited access as well as good ventilation (4).

REFERENCES

1. NFPA (1966). Identification Systems for Fire Hazards of Materials, Publication 704-M-1966. National Fire Protection Association, Quincy, MA.
2. Code of Federal Regulations, 29 CFR 1910.
3. *Federal Register:* 48 FR 53280-53347, November 25, 1983; FR 52 31852-31886, August 24, 1987; 55FR 3300-3335, January 31, 1990.
4. *Prudent Practices for Handling Hazardous Chemicals in Laboratories*, National Research Council, National Academy Press, Washington, DC, 1981.
5. D. D. Perrin and W. L. F. Armarego (1988). *Purification of Laboratory Chemicals* 3d ed., Pergamon, Elmsford, NY.
6. H. L. Jackson, W. B. McCormack, C. S. Rondesvedt, K. C. Smeltz, and I. E. Viele (1970). *J. Chem. Ed.* **47**, A175–A188.
7. J. Varjavandi and O. L. Mageli (1971). *J. Chem. Ed.* **48**, A451–A456.
8. *Aldrich Chemical Catalog*, Aldrich Chemical Company, Milwaukee, WI, 1988.

CHAPTER 6

EMERGENCY RESPONSE FOR RELEASES OF LABORATORY CHEMICALS

DAVID A. PIPITONE
Senior Consultant
Gallagher Bassett Services, Inc.
Rolling Meadows, Illinois

6.1 INTRODUCTION

Releases of chemicals to the environment and within facilities have become an increasingly popular topic in the news media. Newspaper headlines and special television broadcasts tell of tragic and near-tragic events (such as the Bhopal and the *Valdez* oil tanker incidents), with resulting evacuations, injuries, and damage to the environment.

Chemical storerooms and laboratories do not have problems associated with massive releases of hazardous chemicals to the same degree as do transportation accidents. Discharge to the environment may happen on a more limited scale. A pool of liquid from a leaking drum may enter a sewer drain, or spilled chemicals that have been stored outside a building may migrate across property boundaries into the ground. For the most part, releases of laboratory chemicals occur inside buildings.

Chemical storerooms house the ongoing stock requirements of hazardous chemicals for a laboratory and facility. In addition, many storerooms are also used as accumulation and storage sites for hazardous laboratory chemical wastes. Many laboratories store chemicals in sufficient volume for immediate laboratory consumption. Tightening requirements of federal and state regulations for hazard communication, hazardous waste, and training for response teams are placing greater demands on the storeroom manager, laboratory supervisor, and facility administrator. A release of chemicals through leaks, accidental spills, or ruptured containers can pose an immediate threat to the health and life of stockroom and laboratory personnel. Improper containment and cleanup of chemical releases can open the door to further liability: civil lawsuits from affected employees, OSHA investigations of the worksite and resulting fines, and other legal implications.

Chemical releases and spills include the accidental discharge of solid, liquid, and/or gaseous chemicals from their secure storage containers. Not all chemical releases have the potential for emergencies. Many small-quantity spills of nontoxic, nonhazardous materials routinely occur in laboratories. One author has suggested that liquid spills be considered an emergency when the volume of spilled material exceeds one liter (1).

Where emergencies do exist and emergency response is conducted by employees, the OSHA has promulgated specific regulations for worker safety in 29 CFR 1910.120 (2). The OSHA excludes certain situations from its definition of emergency response. Responses to incidental releases of hazardous chemicals where the substance can be controlled at the time of release by employees in the immediate area or by maintenance personnel are not considered emergency responses. In addition, for a situation in which the released chemical has no potential for fire, explosion, chemical exposure, or other safety/health hazard, an effort to control the situation is not an emergency response.

Let us consider two simple scenarios to illustrate the potential danger posed by hazardous chemicals when spilled or released. In Scenario 1, 2.5 liters (L) of a liquid falls from a laboratory bench and breaks on the floor. The liquid forms a puddle, and vapors form and scatter as it evaporates. Let us assume that the vapors spread uniformly into a spherical area with a radius of 2 meters (M) around the puddle. Ignoring dispersion outside the sphere or the effect of ventilation, the more the puddle evaporates, the greater the concentration of vapors around it. Table 6.1 shows the amount of milliliters of liquid that need to evaporate to reach dangerous concentration levels of vapors. Note that in this model only 20 mL of carbon disulfide and carbon tetrachloride would have to evaporate to reach the IDLH (Immediately Dangerous to Life and Health) level.

In Scenario 2, a bottle of a given volume of liquid falls from a shelf in a storeroom and shatters, releasing all the liquid on the floor. In this scenario, the bottle breaks at night when the ventilation is shut off. By the time morning arrives, the liquid is completely evaporated. Table 6.2 shows estimated vapor concentrations in parts per million (ppm) in the storeroom for 250 mL, 500 mL, 1 L, and 4 L of various liquids. For example, evaporation of 250 mL of acetic acid in the specified storeroom would reach the IDLH concentration. (This scenario assumes uniform mixing and ignores the volume of air taken up by shelving, tables, or other objects.)

Either scenario in real life would be far more complicated and offer a greater danger when pockets of vapors form, which are more concentrated. These simple scenarios illustrate the hazard potential for exposure to workers responding to clean up the spill or release. Larger volume releases in smaller areas only magnify the danger to the worker.

There are a multitude of hazardous and toxic chemicals in storage and many instances where an emergency may occur. A spilled hazardous chemi-

TABLE 6.1 Comparison of Health/Safety Values to Spill Evaporation Volumes

Chemical	Concentration (ppm)			Amount (mL) Evaporated to Reach[a]		
	STEL	IDLH	Lower Flammability Limit	STEL	IDLH	LFL
Carbon tetrachloride	25	300	N/A	1.7	20	N/A
Carbon disulfide	100	500	13 000	4.2	20	550
Acetic acid	40	1000	54 000	1.6	40	2150
Chloroform	25	1000	N/A	1.4	60	N/A
Benzene	25	2000	14 000	1.6	120	870
Toluene	150	2000	12 700	11.1	150	940
Acetonitrile	40	4000	44 000	1.5	150	1600
Pyridine	N/A	3600	18 000	N/A	200	1020
Acetaldehyde	50	10 000	39 700	2.0	390	1550
Heptane	500	4250	11 000	59.3	500	1300
Ethyl acetate	1000	10 000	21 800	57.6	580	1260
Methanol	250	25 000	67 200	7.1	710	1910
Xylene	150	10 000	10 000	12.9	860	860
2-Butanol	150	10 000	16 800	15.1	1010	1690
Acetone	1000	20 000	26 000	51.5	1030	1340
Ethyl ether	1000	19 000	18 500	86.2	1640	1600

[a] Assuming that vapors from a 2.5-L spill will occupy 16.8 m^3 of air around the spilled liquid (even distribution).

TABLE 6.2 Comparison of Health/Safety Values to Vapor Accumulation in Storerooms after an Unsuspected Release[a]

	Concentration (ppm)			ppm Resulting from Total Evaporation[b] Volume of Bottle (mL)			
Chemical	STEL	IDLH	Lower Flammability Limit	250	500	1000	4000
Carbon tetrachloride	25	300	N/A	590	1180	2360	9440
Carbon disulfide	100	500	13000	950	1900	3790	15170
Acetic acid	40	1000	54000	1000	2000	4010	16030
Chloroform	25	1000	N/A	710	1420	2840	11360
Benzene	25	2000	14000	640	1290	2570	10290
Toluene	150	2000	12700	540	1070	2150	8590
Acetonitrile	40	4000	44000	1100	2190	4380	17530
Pyridine	N/A	3600	18000	710	1410	2830	11300
Acetaldehyde	50	10000	39700	1020	2040	4090	16360
Heptane	500	4250	11000	340	670	1350	5380
Ethyl acetate	1000	10000	21800	690	1380	2770	11080
Methanol	250	25000	67200	1400	2810	5610	22460
Xylene	150	10000	10000	460	930	1850	7410
2-Butanol	150	10000	16800	400	790	1590	6340
Acetone	1000	20000	26000	780	1550	3100	12400
Ethyl ether	1000	19000	18500	460	930	1850	7400

[a] Room size: 7 long by 5 wide by 3 m high. Assumes equal mixing and no effect by ventilation.
[b] Values rounded to the nearest 10 ppm.

cal represents the danger of exposure to toxic and corrosive material, potential for fire, as well as an excruciating slip on a wet surface. Section 6.3 outlines OSHA requirements for in-house employee response teams and the requirements for coordinating external emergency response teams (such as from the local fire department). Fortunately, planning and implementing an effective emergency release prevention program can reduce the frequency and severity of chemical releases (3).

6.2 PREVENTING CHEMICAL RELEASES

Accidental releases can be avoided by planning and implementing appropriate handling measures and inventory controls. Releases of laboratory chemicals can occur by (a) gradual deterioration or sudden accidental rupture of chemical containers, (b) inadequate shelving space, (c) inadequate shelving integrity, (d) lack of guards on shelves, and (e) inappropriate handling techniques. Using controls, these causes can be eliminated or lessened so that the risk of spills is reduced.

Chemicals can be released by the deterioration of their containers. Containers that have aged, as may be evidenced by rusted metal cans or weakened plastic bottles, lack the structural integrity and holding capability of their original condition. Overpressurized containers, as evidenced by the deformation in container shape, have the potential for rupture and chemical spills. Remedies for leaking and overpressurized 55-gallon drums have been discussed in the literature (4). A regular visual inspection of the chemical stock will reveal those containers that are leaking or have the potential to leak. Repackaging the chemical into another container must be accomplished promptly when suspect containers are identified. If chemical purity is suspect, disposal of both chemicals and their containers is in order.

In storage areas such as docks or warehouses 55-gallon drums can be ruptured by forklifts or other material handling equipment. A secondary containment device, which holds the drum and protects it from accidental impact, can prevent leaks. Other methods of containing leaks include using mesh-covered storage platforms, which will hold any released liquids from a leaking drum.

The shelving system can be the cause of a major chemical spill. Inadequate shelving space that results in overcrowded shelves may well surpass the rated weight limit of the shelving. A prudent measure for remedying this problem is to obtain the weight limit of the shelf (from the shelving vendor) and determine the weight of chemicals (from the containers) that can be placed on the shelf. The weight of bulky containers should be considered for large glass carboys and other items. Keeping a running tally of the accumulated weight on each shelf can indicate how close to the maximum weight limit the current shelf of chemicals is. (The tally can be automated and easily updated using a microcomputer software database, which tracks chemicals by container and location.)

Lack of shelving space encourages the storage of chemical containers on the floor. This practice should be avoided because those containers may be kicked over and start to leak. As a practical measure, larger containers of chemicals, such as 5-gallon drums and carboys, should be stored on shelves close to the floor. Should such a container be dislodged from a low-to-the-floor position, the force of impact when striking the floor is less than if the container is stored at a higher position, thus reducing the chance of a more severe rupture.

Shelving units should be braced or secured to the wall or floor. Accidents have occurred in which shelving supports have buckled when inadvertently jarred by stockroom attendants. In an overstressed condition, entire shelving units may collapse without notice or, even more dangerously, fall upon workers reaching for a chemical. Fastening the shelving units securely and testing the integrity of the secured shelving units before stocking chemicals should be considered.

Random and sustained vibrations in the walls or floors of a room can cause chemicals to creep toward and over the edge of a shelf (5). Providing a raised edge on the end of a shelf can help prevent this migration. Using a simple metal strip that protrudes ¼ inch above the shelf plane is an economical and simple procedure that can prevent an unexpected spill. Installation of a raised edge is demonstrated in Figure 6.1. In areas where extraordinary vibration can exist (like earthquake-prone zones), additional protection, such as the use of bunji straps or enclosed shelving units, should be provided.

Inappropriate handling or transporting of chemicals can result in a ruptured, leaking container. Solutions for safe transport of small containers (4 L or less) include the use of carts, bottle carriers, or pails. These carriers should be in good condition and be able to hold the chemical containers safely, without contributing to a spill situation. The carrying handle and well of bottle carriers and pails should be secure. Cart beds should be strong enough to hold the desired load. Large drums (holding more than 15 gallons) should be transported using a drum truck, pallet jack, or forklift. Some vendors supply specially designed skids to protect drums during transport.

Many chemical suppliers now package chemicals in glass bottles that are encapsulated in plastic. If the bottle is dropped and the glass breaks, the plastic envelope holds the liquid inside. Empty encapsulated bottles are also available commercially to accommodate the repackaging of chemicals when necessary.

Releases of chemicals within laboratories can result from inappropriate dispensing techniques or equipment when the chemicals are transferred from stock containers. Using funnels or pumps can reduce the chance of "sloshing" liquids onto countertops or floors. Operations should pay attention to the rate of dispensing flow and liquid level in the transfer container to avoid overfills and spills. Placing a spill control tray under a stationary container at a dispensing station can help collect drips, overfills, and unanticipated discharges.

Smooth operating procedures, material handling, and scrutiny of the storeroom contents are the best management controls for spill prevention. Establish-

FIGURE 6.1 Example of a raised lip on the edge of a shelf to keep bottles from migrating over the edge. (Used with permission of Lab Safety Supply Company, Janesville, WI.)

ing a routine visual inspection on a regular basis can result in the reduction of accidental spillage. The checklist in Table 6.3 may be adopted for this purpose.

6.3 PLANNING FOR EMERGENCY RESPONSE

The previous section discussed the possible causes of chemical releases or spills and proactive ways to eliminate those causes. Emergency response seeks to minimize the effects of a spilled chemical by acting quickly to clean up a spill.

Releases of laboratory chemicals can occur during the transport from the receiving dock, during storage and handling within a storeroom or laboratory,

TABLE 6.3   Inspection Checklist for Identifying Causes of Spills

	Yes	No[a]
Metal containers are free of rust.		
Containers are clean and free of any chemical leakage.		
Container shapes are normal, free of any signs of presure buildup.		
Shelving units are fastened to the wall and/or floor.		
Shelves are free of overcrowding.		
Weight limiits for shelves are posted on the shelving units.		
Weight limits for shelves are not exceeded by the weight of chemicals.		
Large containers are stored on lower shelves.		
Raised edges are an integral part of each shelf.		
Shelving units are braced.		
Carts, bottle carriers, or pails are used to transport chemical containers.		
Carts, bottle carriers, or pails are in good physical condition.		
Funnels or pumps are used to dispense or transfer liquids.		
Spill control trays or basins are used to collect overfills at dispensing stations.		

[a] *No* answers indicate a deficiency that can lead to a chemical release through a leak or spill.

and during dispensing operations. These situations represent the potential for uncontrolled release and discharge of a chemical.

Effective response depends on the knowledge of the spilled material, accurate choice of response equipment and gear, and trained personnel to use established procedures to clean up the area and dispose of the spent material in a proper fashion. An Emergency Response Plan is a written, structured set of procedures, methods, and techniques which communicate how to control laboratory chemicals that have been released to the workplace environment. Such a plan serves as a guide for personnel responding to the scene of a chemical release.

The possession and use of an Emergency Response Plan for any facility storing laboratory chemicals is essential to inform stockroom or laboratory personnel and emergency response team members of response strategies.

This section discusses emergency response and is limited to releases of laboratory chemicals occurring within a building. Discussion of massive releases of chemicals, such as from tanks or piping systems with volumes exceeding 100 gallons, is beyond the scope of this chapter.

Some chemicals pose minimal hazards, if any at all, when accidentally released from their containers. The real danger exists when personnel encounter an unknown puddle or pile of material on the floor. There are a variety of hazardous chemicals (Table 6.4) that may require different response tactics arising from the nature of the chemicals and their individual hazards. Lee (6) indicates five areas of concern when a spill confronts personnel: (a) lack of information about the hazard, (b) lack of information about resources, (c) lack of tactical information, (d) lack of response capability, and (e) lack of training.

Because a chemical release often occurs unexpectedly, having an informed plan is the first measure for emergency response. For the purposes of emergency response to releases of laboratory chemicals, OSHA specifies two distinct sets of requirements. For releases of hazardous laboratory waste at a hazardous waste storage facility (see Chapter 2 for EPA requirements), facilities that have their own emergency response teams must comply with 29 CFR 1910.120(p). For all other emergency level releases of laboratory chemicals, facilities must comply with 29 CFR 1910.120(q). Table 6.5 shows the differences in requirements for the two types of facilities.

A useful plan will (a) contain evacuation and crisis management procedures, (b) inform response personnel of hazards and current resources for response, (c) provide written procedures and tactics for response, and (d) schedule and maintain training of personnel.

The depth of an Emergency Response Plan may vary widely. Developing and implementing the plan will depend on the responsiveness and ability of those attending the storeroom, the complexity of the facility's operations, the size of the chemical inventory, and the nature of the chemicals released. For example, a school with a small closet as a chemical storeroom would have limited funds and personnel to respond to spills. The science department would be well advised to rely on a trained, well-equipped response team from a local fire department for spills of toxic substances or large releases of flammable liquids. In practice, this means the science department should supply a copy of Material Safety Data Sheets (MSDSs) and other hazard information to the local emergency response team before a release occurs. A large institu-

TABLE 6.4 Examples of Laboratory Chemicals for Which an Emergency Release Response Plan May Be Necessary

Liquids	Solids
Acids	Oxidizers
Mineral	Water-reactives
Oxidizing	Pyrophorics
Organic	Caustics
Caustics	Toxic acids
Flammables	
Toxic reducing agents	*Gases (Compressed)*
Oxidizers	
Mercury	Toxic
Organic peroxides	Reactive
Water-reactives	Flammable
Pyrophorics	

TABLE 6.5 List of OSHA Requirements for Employee Emergency Response Teams

Requirements	Hazardous Waste Storage Facility	All Other Laboratory Chemicals
Safety and Health Program	X	
Hazard Communication Program	X	
Medical Surveillance Program	X	X
Decontamination Program	X	
New Technology Program	X	
Material Handling Program	X	
Training Program	X	X
Emergency Response Program	X	
Emergency response plan	X	X
Preemergency planning	X	X
Personnel roles	X	X
Emergency recognition and prevention	X	X
Safe distances/refuge	X	X
Site security/control	X	X
Evacuation routes	X	X
Decontamination	X	X
Emergency medical treatment	X	X
Emergency alert/response procedures	X	X
Critique of response	X	X
PPE and emergency equipment	X	X
Procedures for handling emergency incidents	X	X
Postemergency response procedures		X

tion, on the other hand, may have an organized response team ready promptly on notification. Whatever the case, the safety of people and the timely control of the release are the major concerns.

6.3.1 Components of an Emergency Response Plan

An Emergency Response Plan based on OSHA requirements must contain the elements listed in Table 6.5. The plan must be in writing and available for inspection and copying by employees, their representatives, and OSHA personnel. If employers evacuate their employees from the workplace when an emergency occurs and do not permit them to assist in emergency response operations, OSHA does not require an Emergency Response Plan. In such a situation, there must be an emergency action plan in place in accordance with 29 CFR 1910.38(a). (See Chapter 2 for a summary of Subparagraph 1910.38(a).)

The following sections describe the details of OSHA requirements for an Emergency Response Plan.

6.3.1.1 *Pre-emergency Planning and Coordination with Outside Parties*

The plan should be developed to anticipate the different types of chemical emergencies that may call for emergency response. This can include identification of areas where releases may occur, identification of the conditions that initiate the release, and potential resulting severity of the release. If a facility expects assistance from the outside, such as from the local fire department or another local emergency response team, these groups should be contacted to make arrangements and specify the conditions for their involvement in emergency response situations.

6.3.1.2 *Personnel, Lines of Authority, Training, and Communication*

The plan must contain information on who is involved in the response effort, what the reporting structure is during response efforts, training, and communication requirements. OSHA has adopted the concept of an incident command system which is site-specific for each response (7). Responsibilities of the response team members is discussed later in Section 6.6. Typical lines of authority at a release site are depicted in Figure 6.2.

6.3.1.3 *Emergency Recognition and Prevention*   As an Emergency Response Plan element, there must be provisions for recognizing and preventing emergencies. Section 6.2 outlined several alternatives for preventing releases of hazardous chemicals that could lead to emergencies. In addition, operations personnel involved with the storage of laboratory chemicals should have the ability to report releases to the proper authority in the facility. First responders can help suppress a more severe emergency by employing prevention techniques.

6.3.1.4 *Safe Distances and Places of Refuge*   The Emergency Response Plan must contain the minimum safe distance from the scene of the emergency release for operating and response personnel. In addition, designated places of refuge must be predetermined or provisions must be made to reassign those places during the emergency.

6.3.1.5 *Site Security and Control*   The Emergency Response Plan must contain provisions for restricting access and denying traffic of unauthorized personnel to the release site. This may include barricading or other physical means to block intruders or curious onlookers.

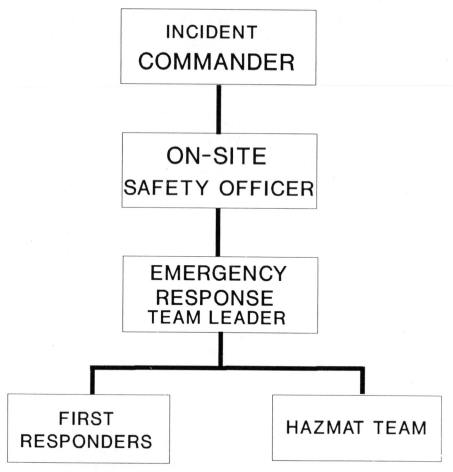

FIGURE 6.2 General lines of authority at emergency response operations.

6.3.1.6 Evacuation Routes and Procedures The Emergency Response Plan must include maps and/or descriptions of evacuation routes, along with written evacuation procedures for use when a release of hazardous chemicals occurs. These routes should be established based on an analysis of the potential locations for chemical releases. Procedures should be included to identify alternative evacuation routes and to direct traffic toward them in the event that primary evacuation routes are not accessible.

6.3.1.7 Decontamination Written methods and procedures for decontaminating personnel, the release area, affected equipment, containers, supplies, clothing, and other items are to be a part of the Emergency Response Plan.

6.3.1.8 Emergency Medical Treatment/First Aid   The plan must include first aid and medical treatment information in the event of an injury or chemical exposure. This information should be easily retrieved and available to the response team. Provisions should be made for medical treatment resources and first aid supplies.

6.3.1.9 Emergency Alerting and Response Procedures   The plan must contain procedures for alerting nonessential response personnel for evacuation, and summoning the emergency response team to the site of the release. Communication methods should be established and included in the plan, which may specify the type of alerting (e.g., siren, alarm, or public address announcement) and summoning (e.g., siren, alarm, telephone contact, etc.).

Definite emergency response procedures must be prescribed and included in the plan. Tactics will vary on the type of chemical and release (these are discussed in more detail in Section 6.4).

6.3.1.10 Critique of Response and Follow-up   The plan must include provisions for evaluating the emergency response operations and incorporating remedies for any deficiencies into the training or into the plan itself. Having a preestablished checklist or evaluation form to compare actual operations to the written guidelines is a convenient way to monitor how effective response operations are.

6.3.1.11 Personal Protective Equipment and Emergency Equipment
The plan must identify and specify the types of personal protective equipment (PPE) and emergency equipment/supplies to be used by the emergency response during its operations. A more detailed discussion of emergency equipment can be found in Section 6.4 and of PPE in Section 6.5.

6.3.2 Emergency Response Data

Emergency response data must include the hazards of the chemical, the type of equipment available for cleanup and its location, procedures for response and cleanup, personal protective equipment, first aid information, and waste disposal procedures. Hazards of the chemical include toxicity, flammability, reactivity with air or water, corrosivity, and its physical characteristics (such as density, state, vapor density, and vapor pressure, where appropriate). For instance, liquid releases can be more dangerous than those of solids because liquids can flow under shelves and doors and across aisles while discharging vapors, mists, or gases.

An incidental step to developing an Emergency Response Plan is to obtain and organize information on each chemical's hazards. Fortunately, there are many good sources of hazard information available in a variety of media (8–

24). The *CHRIS Hazardous Chemical Data* (8) developed for the U.S. Coast Guard for response to water spills of hazardous materials and the *NIOSH–OSHA Pocket Book to Chemical Hazards* (11) give summaries of health, flammability, and reactivity hazards for a large number of chemicals. *Toxic and Hazardous Chemicals in Industry* (10), *Handling Chemicals Safely* (9), and *Dangerous Properties of Industrial Materials* (12) also provide chemical hazard and handling data. Hazard information is also available electronically through mainframe and microcomputer databases. Manufacturers and suppliers of laboratory chemicals may make special hazard information databases available to their customers.

To have a complete system of information on all the hazardous chemicals in stock, designated persons must take an inventory of those chemicals, note which chemicals require special handling, and develop an appropriate data file for response purposes. Gathering inventory data may coincide with the requirements of SARA Title III activities for the facility. There are a number of commercially available computer database packages that can facilitate the inventory process.

The type and amount of cleanup equipment to be made available will depend on the type and amount of chemicals that could be released. A release of chemicals in solid form requires different response procedures than those for liquids. Releases of corrosive liquid acids may be handled differently than those of flammable liquids. The physical size of the spill affects the amount and type of absorbents, neutralizers, or mechanical equipment needed for response. A release of toxic chemicals may require the use of different apparatus. Section 6.5 discusses the use of such equipment at length.

Procedures for cleanup will depend on the location and nature of the release. Procedures for neutralization differ from those of absorption. Procedures for picking up a solid differ from those for a liquid. Containing a release of compressed gas uses a different procedure than preventing the spread of a liquid release. For liquid chemicals with differing concentrations, cleanup procedures should reflect the appropriate level of caution, based on the degree of chemical hazard. A concentrated acid will be more dangerous to work with than a dilute one. Flammable liquids require more care than combustible liquids, and so on. A written procedure that can be retrieved as an instruction sheet at the time of an emergency will do much to guide the response personnel.

Personal protective equipment must be donned before entering the scene of a release. Responders should know what kind of equipment should be worn for protection against a given chemical. A plan based on research will identify the appropriate protective clothing, eyewear, gloves, and respiratory protection required for handling such an emergency. The amount of protective gear required will again depend on spill size, method of treatment (e.g., neutralization of an acid will cause splattering of that acid), and the presence and degree of the hazardous properties (e.g., toxicity, corrosivity, etc.).

First aid information is important in the event of an injury or chemical exposure. The possession of informed manuals (10,16,24) with detailed first aid procedures for each chemical is a must. It is essential that this information be easily retrieved by response personnel. Additional information and supplies for medical treatment should be readily available.

Information for packaging the cleanup debris and waste disposal procedures must be known so that any resulting hazardous material is properly containerized and promptly discarded. Awareness of federal (25) and state regulations for the proper disposal of hazardous waste [including chemical residues or debris of those spent chemicals listed in 40 CFR, Section 261.33 (e), (f)] is essential. Since cleanup debris will be stored until transported from the facility, spent cleanup supplies should be packaged in leakproof containers to avert further leaks or releases.

A designated person should develop, implement, and monitor the Emergency Response Plan for chemical releases. This person should have the authority and ability to dispatch and manage response personnel in the event of an emergency. The authorized person may be a chemical safety officer, storeroom manager, or laboratory supervisor, depending on the organizational structure of the facility.

The preceding discussion points to the development and management of a systematic information system that can supply procedures and facts to personnel responding to chemical releases. The complexity of the information system depends on the need, resources, and inventiveness of the designated chemical safety officer for the organization. For some, a reference library may suffice. For many organizations, a useful chemical information system could be developed using microcomputer database software to accelerate the retrieval process upon demand of the responders.

Table 6.6 shows the functional elements of such a conceptual database. This format mimics an MSDS; however, the reader should note specific references to the location of the stored chemicals, recommended type of spill response equipment, an exacting spill response procedure, and so on. The content of the information system should be designed for quick reference and extraction of the needed information. The database must be practical and reliable to work effectively. If a microcomputer database houses the emergency response information, the software should be menu driven and include fast query capabilities. More information on computer systems for emergency response may be found in reference 26.

An Emergency Response Plan should be integrated into the overall disaster plans of an organization. Provision for evacuation, communication, crisis management controls, and backup support should be addressed within the overall framework of the disaster plans.

The remainder of this chapter discusses the selection and use of those functional components that can "fill the data blanks" when developing the Emergency Response Plan.

TABLE 6.6 Conceptual Response Plan for a Spill of Acetone

Chemical: Acetone
Chemical class: Flammable liquid

		Number of containers			
		500 mL	Liter	4 L	20 L
Location:	Room B1297	5	0	10	2
	Room B1298	7	5	15	5
	Room B2397	0	0	20	10

Chemical Hazard Data

Health: TWA, 750 ppm; PEL, 1000 ppm; IDLH, 20,000 ppm. Avoid skin contact.

Flammability: Flash point: 0°F (CC); LFL 2.6%; UFL 12.8% Flashback may occur.

Spill Response Equipment/Location

1. Absorbent pillows, Central Stores, Closet 2. Use one pillow per gallon spilled.
2. Diatomaceous earth. Room B1263. Spread until liquid is absorobed.
3. Nonsparking shovel. Room B1263. Use to pick up spent absorbent.
4. Empty 17H drums. Room B1263.
5. Barricade tape, warning signs. Room B1263.

Protective Clothing/Location

1. Disposable jumpsuits, Central Stores, Closet 1, Shelf 1.
2. Butyl rubber gloves, Central Stores, Closet 1, Shelf 2.
3. Self-contained breathing apparatus, Central Stores, Row 1.

Fire Response

Use carbon dioxide, ABC dry chemical, or Halon 1211 fire extinguisher on small flames.

Spill Response Procedure

1. Barricade spill scene from traffic and evacuate nonessential personnel.
2. Notify safety officer.
3. Contact response team: Bill (Ext. 3242), Joel (Ext. 3246).
4. Measure vapor concentration if appropriate.
5. Review chemical hazards and don appropriate equipment.
6. Obtain appropriate spill response equipment.
7. Remove all ignition sources.
8. Dike spill and apply absorbent to spilled liquid.
9. Shovel debris into drum and secure cover.
10. Scrub and ventilate spill area.
11. Seal drum and label with contents.
12. Report spill cleanup details.
13. Replenish spill response equipment.

TABLE 6.6   (*Continued*)

First Aid

Inhalation: Remove to fresh air. If breathing has stopped, give artificial respiration. If breathing is difficult, administer oxygen.

Eye exposure: Flush eyes with lukewarm, potable water for a minimum of 15 minutes, holding eyelids open.

Skin exposure: Flush affected area of skin with lukewarm, potable water for a minimum of 15 minutes.

Waste Disposal

Spill debris is EPA hazardous waste (D0001-ignitable).

DOT label: FLAMMABLE SOLID.

Place debris in DOT Spec 17H metal drum. Send drum to approved incinerator or treatment facility.

6.3.3   Chemical Hazard Information

The need for specific information is the first consideration when responding to the release of a particular chemical. The response team can be exposed to hazardous chemicals at toxic or hazardous concentrations. In developing a response plan, chemical hazards must be identified and communicated to the response team. A summary of hazards, with references, follows.

Health hazards   Inhalation: TLV, IDHL, LC_{50} (9-12,26); skin exposure: TWA, LC_{50} (10,27); corrosivity to skin (9,10); absorption through skin (8,10); toxicity by ingestion (8,10)

Fire hazards   Flash point (8-11), special hazards of combustion products (8,10), upper/lower flammability limits (8-10), type of extinguisher needed (8-17), potential for flashback (8), ignition temperature (8-11)

Reactivity hazards   Water-reactive (8-15,21), pyrophoric (5,8-15,21)

Physicochemical properties   Physical state at 20 °C (10-22), vapor (gas) density (9-14), specific gravity (8-12), evaporation rate (28)

This information should be obtained and recorded for each hazardous laboratory chemical stored in a facility. A hazardous material can generally fall into one or more of the following categories (29):

Acid   Water-reactive chemicals
Caustic   Pyrophoric chemicals

Flammable liquids   Poisons (solid or liquid)
Oxidizers   Other reactive chemicals

An inventory list may be compared against these general hazard classes to obtain the names, locations, and amounts of hazardous chemicals in stock.

The use of MSDSs (OSHA 20 form) provided by a supplier is an easier route than compiling hazard data from the listed references. Several commercial software vendors provide a database of chemicals that also include hazard data for specific chemicals. A practical approach to obtaining hazard data is to obtain all MSDSs from suppliers at purchase and examine the completeness of the data. Many data sheets have a limited amount of hazard information and are of little use in compiling emergency response data. The use of a mediocre MSDS for response data can leave knowledge gaps, which are undesirable in an emergency situation. This can represent an opportunity for the buyer of laboratory chemicals to put pressure on the chemical vendor to supply complete data. Purchase orders can specify that complete MSDSs must accompany the chemical shipment. Received shipments that do not meet that criterion will be refused. If your vendors do not respond, you may consider seeking out another supplier or augmenting the MSDS information with literature-based procedures, emergency equipment manufacturer's directions, and other hazard data.

Purchasing a "canned" chemical response plan may work in some instances where the types of chemicals are standardized in amount and type. This may be feasible for high school laboratories that are part of a school district insurance pool and share limited laboratory experiments in the science. Large-volume, analytical laboratories that run highly automated, specialized analyses may store only a limited variety of hazardous chemicals. For many research, college, industrial, and analytical laboratories, the number, amount, and diversity of chemicals literally require a customized emergency response database.

6.4 RESPONSE PROCEDURES FOR CONTAINING CHEMICAL RELEASES

The first determination that must be made for releases of laboratory chemicals is whether the release is, or has the potential for, an emergency. If hazardous chemicals are involved that present the potential for safety and health hazards, and personnel in the immediate vicinity are unable to respond, two options exist. The organization may evacuate the site and call for help (if there is no emergency response team), or it may activate the emergency response team. Section 6.4.1 lists the general procedures required by the OSHA if an emergency response team is activated to respond to the release. The following sections discuss general methods for containing different types of chemical releases.

6.4.1 General Emergency Response Procedures (OSHA)

If an emergency response team responds to an emergency release (as defined by OSHA's 1910.120 standard), there are specific procedures for handling the emergency response, as follows:

1. The senior emergency response official responding to a release shall become in charge of the Incident Command System. This individual will coordinate and control all emergency responders and their communications.
2. The Incident Commander shall assess the situation, identify the hazardous chemicals involved and the conditions of the release, and address the use of handling procedures, engineering controls, and maximum exposure limits.
3. The Incident Commander shall implement the appropriate emergency operations and direct the emergency response team to wear appropriate personal protective equipment. (See Section 6.5 for additional details.)
4. The Incident Commander will limit the number of emergency response personnel in the emergency site to those actively performing the emergency operations. Such operations must be performed using the "buddy system" in groups of two or more.
5. The Incident Commander will designate a safety official to identify and evaluate hazards threatening the safety and health of the response team and the operations at hand. When conditions reach the IDLH condition or an imminent danger condition, the safety official will have the authority to alter, suspend, or terminate response activities, informing the Incident Commander of any actions needed to correct such hazards.
6. Any emergency response employees who exhibit signs or symptoms of chemical exposure shall be provided with medical consultation, either immediately or subsequently to the operations.
7. After emergency operations have ended, the Incident Commander shall implement appropriate decontamination procedures.

6.4.2 Considerations for Solid Spills

Releases of solid chemicals are easier to contain than those involving liquids for three reasons: (a) solid chemicals are normally packaged in smaller containers; (b) spills of solid chemicals do not flow uncontrollably as do liquid spills; and (c) most solid spills can be physically removed from the floor or shelf without much special equipment. Spills of solid hazardous materials can, however, demand more careful attention because of their highly concentrated form and sometimes toxic, flammable, or reactive nature. When spilled, solids can create aerosols of microscopic dusts, which may unknowingly be inhaled by personnel in the vicinity. Spills of many solid chemicals can normally be swept by a broom into a dust pan and placed in a suitable waste container, constructed of the same materials as the original container. Spilled oxidizing solids, such as nitrates, permanganates, and perchlorates, should be separated from other types of waste products and should be especially kept away from combustible materials, such as paper. Spills of extremely toxic solids that yield toxic dusts (e.g., beryllium, cadmium, arsenic compounds, barium, or mercury compounds) are frequently collected using a HEPA-filtered vacuum cleaner (30). Such a vacuum cleaner has an absolute filter that removes up to 99.97% of particles that have a mean diameter as small as 0.3 μm.

A spill of white phosphorus evokes danger because it burns when exposed to air. Keeping spilled phosphorous wet and then covering it with wet sand is a preliminary action before recovering the material (31). The spill residue must be kept underwater to prevent reignition. Other pyrophoric materials must be investigated for their compatibility with water before using this method.

Spills of water-reactive solids such as sodium or potassium must be treated completely differently. These chemicals react with water to form flammable hydrogen gas that can be ignited by the heat of the reaction. Covering potassium with dry sodium carbonate and dispersing and incinerating the mixture in a large steel pan located in an isolated place is recommended by one source (32). Sodium and potassium are normally stored under mineral oil to prevent contact with moist air. Recovered material should be placed into a crock containing enough mineral oil to cover the metal.

Spills of explosive solids present a high degree of danger. These spills are normally saturated with a liquid to desensitize the material. Sodium sulfate desensitizes most but not all explosives (33). Only trained, authorized personnel should attempt to clean up a spill of explosives.

6.4.3 Considerations for Releases of Hazardous Liquids

Releases of hazardous liquids can be dangerous for the following reasons: (a) liquids can flow to other areas, bringing the hazardous properties along; (b) liquids can emit gases or vapors that can be toxic, flammable, or corrosive; and (c) liquids present a greater slipping hazard than do solids.

With the advent of EPA hazardous waste regulations, diluting liquid laboratory chemicals with water for the sole purpose of eliminating hazards is no longer an efficient emergency response strategy. Dilution may seem an effective way to reduce the hazardous nature of concentrated corrosives, such as acids; however, the volume of water needed, and the amount of subsequent hazardous waste generated, can drive up the ultimate cost of disposal. If a spilled liquid is water-reactive (such as concentrated sulfuric acid), dilution with water in situ (i.e., on the floor) can cause splattering and related harm. Dilution of a liquid with water to make it less hazardous will not work for organic solvents and other hazardous liquid spills. Flushing a chemical release into the sewer system is no longer conscionable or legal. The liquid must be removed from the environment in a proper fashion.

The extent of the emergency involved with releases of liquid chemicals is time-driven. This depends on the potential volume of liquid that can be released, the type of incident that causes the chemical release, the rate of release, and when the release is discovered and contained. For example, within a minute or two after a 2.5-liter bottle of nitric acid is dropped, the amount of chemical released has stabilized and accumulated into a pool of liquid. The emergency response procedures would be limited to diking the perimeter of the pool and absorbing the liquid. On the other hand, a punctured drum will continue to spout liquid until the liquid level falls below the rupture. In this case, the chemical release is dynamic and should be handled as such. The boundary of the liquid pool should be diked to prevent further spread, the drum leak plugged or patched using an approved response kit (34), and the released liquid absorbed. The state of the release, whether dynamic or static, will determine the type of response methods and emergency equipment used.

6.4.3.1 *Controlling Vapors*

The hazards presented by the vapors of liquids can be as dangerous as the released chemical itself. Releases of flammable liquids give off flammable vapors that can travel to an ignition source and flashback to the liquid pool. Corrosive and toxic gases can cause injury to unsuspecting personnel, especially if allowed to concentrate in a small area. Preventing the spread of vapors, gases, and mists is a vital part of the response effort.

Attention must be called to the saturated vapor concentration. Releases and spills occurring inside a building can concentrate vapors in an area around and over the spill of a volatile material. This condition represents a hidden danger. A pocket of toxic or corrosive gases or vapors can expose unsuspecting personnel to dangerous concentrations. Pitt (35), in discussing a vapor hazard index, calculates the relative hazard involved with the accumulation of vapors by comparing the ratio of saturated vapor concentration to the threshold limit value. (The reader may wish to refer back to Tables 6.1 and 6.2, which approximate vapor concentrations for liquid spills and storeroom releases for several hazardous liquids.)

When spills occur in an unventilated area, these concentrations indicate the degree of hazard that can develop over time. The vapor concentration in the breathing zone near a recently spilled liquid will depend on the rate of evaporation, the effect of ventilation, and the time elapsed since the spill occurred. These values are particularly important when determining what type of respiratory protective equipment should be worn.

Providing ventilation paths for the vapors to fume hoods or fume scrubbers can help reduce the vapor concentration around a spill. Using fans or other local exhaust ventilation may work for some releases, but should not be used for vapors from extremely volatile flammable liquids with a low flash point. Vapors may also be suppressed by covering the spilled liquid with a special foam or impregnated activated charcoal. This results in blanketing the liquid to keep the vapors from escaping.

Liquids cannot be removed from a surface easily without wicking or converting them into a solid state. This may be done by adding an absorbent or gelling agent directly to the spilled liquid. There are two main approaches of emergency response for most liquid chemicals: (a) inactivate the hazardous properties of the spilled chemical immediately and then remove it from the environment (neutralization); and (b) remove the spilled chemical (hazards and all) from the environment immediately and treat it later (absorption of untreated liquids). Before choosing either method, the reader must understand which method works better for both the chemical and the situation.

6.4.3.2 Chemical Inactivation (Neutralization)

Chemical inactivation depends on the interaction of two chemicals to result in a harmless third chemical that may be easily recovered and discarded. Commonly referred to as neutralization, this method has been traditionally used for response to spills of inorganic acids and bases, although it may be applied to other chemicals (36). Addition of the proper amounts of a weak base (e.g., sodium bicarbonate, sodium carbonate, or calcium carbonate) to strong acids, and weak acids (e.g., citric acid) to strong bases (e.g., hydroxide solutions) results in a chemically neutral solution of salt and water. The use of concentrated neutralizing agents (e.g., sodium hydroxide pellets for acids) is not recommended because dangerous splattering can occur.

Neutralizing spilled acids and bases requires using the correct amount of neutralizing agent. Since the volume and molarity of acids and caustics differ, determining the proper amount of neutralizer to add to the spilled material is not an easy task. Neutralizing acids using a carbonate—bicarbonate mixture produces gaseous carbon dioxide. The lack of foaming or fizzing signals the endpoint of neutralization. A drawback of this measure is the uncertainty of complete mixing of the spilled material with the neutralizer. Checking the mixture with pH paper or pH indicator dye will give a more accurate indication of neutrality. The use of citric acid on caustic spills, on the other hand, does not show a neutralization endpoint by foaming. The pH indicator test is essential in this case.

Since neutralization is a strong exothermic reaction, the generated heat can cause splattering of the spilled material. Concentrated acids and caustics are found in the majority of laboratories and stockrooms. The addition of neutralizers to releases of those chemicals at a fast rate will cause violent splattering. Neutralization of these chemicals is necessarily a slow procedure. For very large spills (one or more gallons), neutralization can be a costly, time-consuming practice.

Several manufacturers of spill kits for neutralizing acids and caustics have prepared prepackaged amounts of neutralizing agents. J. T. Baker Inc. has formulated neutralizing products that produce a color change to indicate a neutralized acid or caustic spill. Such kits are beneficial in that the rated capabilities are given for small spills. However, one must evaluate the economics of using them as well as inventorying bulk quantities needed to neutralize large spills.

Hydrofluoric acid deserves special mention because of its extreme corrosivity to human tissue. The acidic fluoride ion attacks skin quickly and without initial pain. The resulting burns are hideously painful and slow to heal. A spill of hydrofluoric acid should be treated with a calcium-containing compound and soda ash to precipitate the fluoride ion as harmless calcium fluoride and render a neutral pH (37). Extra care must be taken to avoid skin contact with the acid, as well as inhalation of hydrogen fluoride gas.

As previously discussed in the section on spills of solid explosives, spills of liquid explosives should be treated in a similar manner, by first desensitizing the material using the appropriate agents.

In summary, neutralization is accomplished by adding a neutralizing agent to a release of strong acid or base. For small spills of acids or caustics, neutralization is a viable method. After neutralization is complete, a wet slurry still remains in the spill area that must be removed by absorption, shoveling, vacuuming, or other mechanical activity.

6.4.3.3 Absorbing Liquid Releases

Some liquids cannot be effectively treated or neutralized before being physically removed from a surface (e.g., organic solvents). The primary response option in these situations is to use an agent to convert the liquid into a semisolid or solidified state. Traditional absorbent methods include the use of mops, gelling agents, and loose or bundled inert absorbents.

Mops should *never* be used to clean up spills of hazardous, toxic, or reactive liquids, for several apparent reasons. Most mop heads are made of cotton strands that can be easily degraded by strong acids or oxidizers, causing an additional mess. If water is added to dilute a spill of a strong acid so that a mop may be used, care must be taken to assess the corrosivity of the final mixture to prevent degradation of a metal mop bucket and ringer. Wringing a mop full of recovered flammable liquids could cause a fire in the event that a spark is generated from the wringing mechanism. Liquids recovered by mopping are normally contaminated and must be discarded properly. A mop, wringer, and bucket should *not* be used on a spill of a hazardous liquid, but may be acceptable and economical for nontoxic, noncorrosive, nonflammable, and inert liquids.

A common method for cleaning up liquid spills is the use of absorbents or gels to convert a spilled liquid into a slurry or solid form. The residue is then placed into a waste container and secured from further releases. Absorbents vary immensely in physical characteristics, absorption capacity, inertness, and suitability for spills. Brugger gives a comprehensive listing of various vegetable (e.g., paper towels, sawdust, corn cobs), mineral (e.g., clay) and synthetic (e.g., polypropylene fibers) absorbents (38). A compatibility table of absorbents for various chemical categories is listed in reference 8. Selection of a universal absorbent that will not break down or lose efficiency when in contact with all types of chemicals is highly desirable.

The more inert materials used to absorb hazardous liquids are mineral in nature. This category includes exfoliated vermiculite, diatomaceous earth, sand, granular clay, and amorphous silicate. Because silicon is a major portion of these absorbents' chemical composition, these absorbents would react with hydrofluoric acid to produce silicon tetrafluoride, a toxic gas. Special synthetic absorbents have been designed for picking up spills of hydrofluoric acid. However, these synthetic absorbents may be degraded by strong oxidizers. All absorbent materials vary in cost-effectiveness and sorption capacity for different liquids.

If a facility plans to use absorption as the primary method of counteracting liquid releases, it is essential to keep enough absorbent on hand in anticipation of response to worst-case chemical releases. Several manufacturers have packaged a predetermined quantity of absorbent in "pillows" that are rated to absorb a given quantity of spilled liquid (such as one gallon or 4 L). By estimating the size of the most likely releases and identifying where those releases can occur, a sufficient quantity of absorbent can be inventoried for anticipated needs.

One important feature of absorbents is the quickness with which the liquid is absorbed. The sooner the liquid is absorbed, the sooner a spill crisis is under control. The speed of absorption by some of the absorbents in pillows is quite rapid. Some manufacturers' literature claims that 98% of the rated capacity of the absorbent in the pillow is reached within 30 seconds. The containment of absorbent within a pillow structure also eliminates the messy step of transferring wet, loose absorbent into a secondary containment device (like a drum).

The volume and weight efficiency of an absorbent is important when considering waste disposal of spent material. The U.S. Environmental Protection Agency (EPA) (39) and many states have published legal requirements for the packaging and disposal of hazardous waste, which includes the contaminated debris and residue from hazardous chemical cleanup operations. Absorbents that absorb a lesser amount of liquids will take up a larger volume when containerized compared to absorbents with higher absorption capacity. This will increase disposal costs because more containers (normally 55-gallon drums) will be needed. Dense absorbents, which have a higher unit weight, can cause a small-quantity generator to exceed EPA specified limits and require compliance with stricter requirements for waste disposal (40).

Using absorbents on a spilled liquids is a straightforward practice. The general procedure for using absorbents on liquids is (a) dike the circumference of the liquid to prevent its spread (by surrounding it with a barrier of absorbent); (b) add absorbent to the liquid; (c) mechanically mix the absorbent with the liquid until a homogenous solid results; (d) place the mixture (normally by shoveling) into a chemically resistant disposal container; and (e) decontaminate the spill area. Note that step d requires knowledge of the corrosivity, reactivity, or flammability of the spilled liquid. Placing an untreated, absorbed acid into a metal drum can corrode the drum and instigate further releases. Absorbents do not change the chemical properties of a hazardous liquid. Absorbed corrosive acids will still be corrosive. Absorbed flammable liquids will still burn when ignited. The main advantage of absorption is the speed of cleanup; however, the need for determining the inertness of the absorbent to the liquid can be this method's main drawback.

6.4.3.4 Mercury Spills Mercury is the only liquid metal listed in the periodic table. Mercury is extremely toxic through ingestion and skin contact. Spilled mercury exposes personnel to the hidden danger of inhaling mercury vapors. Mercury vapors have no odor or readily discernible warning properties. Mercury's PEL by inhalation is 0.05 mg/m^3. Mercury is very mobile and as such can easily fill cracks and crevices in benchtops, floors, and other surfaces. Without sophisticated vapor monitoring equipment, spilled liquid mercury can completely evaporate without anyone being aware that toxic vapor is present.

Traditional methods of cleaning up mercury spills have included sprinkling powdered sulfur over the mercury-contaminated floor to precipitate mercuric sulfide on the surface of the mercury, thus inhibiting evaporation. More recent methods include the application of amalgamating powders to bind metallic mercury into an amalgam and preventing vapor formation. Foam collectors have been used to recover small droplets of mercury. Specially impregnated granular activated charcoal has been used to cover suspected areas of mercury contamination so as to absorb the vapors and prevent contamination of room air.

Storerooms and laboratories that store kilogram quantities of mercury as well as scientific apparatus such as manometers should purchase and use a specially designed vacuum cleaner. Figure 6.3 shows one type of vacuum cleaner that incorporates a device for separating liquid mercury from the rest of spill debris by centrifugal action. The mercury is recovered and deposited in a separate reservoir. Such a vacuum cleaner is equipped with special charcoal filters to scrub the vacuum exhaust air of mercury vapors. Final assessment of mercury cleanup and decontamination should be achieved using a special mercury vapor monitor or impregnated indicator paper, which changes color in the presence of mercury vapor.

6.4.3.5 Releases of Cryogenic Liquids Cryogenic liquids, liquids with a boiling point below $-238\,°F$, pose special hazards when a spill or release occurs. When released from closed pressure systems to a room temperature environment, cryogens boil vigorously to produce very large volumes of gas. The gases from most cryogenic liquids have no sensory properties, such as taste, smell, or color. Releases of cryogens can be detected by the appearance of a heavy fog (condensation of moisture) which appears over the spilled liquid. The excess of these gases can create an oxygen deficiency quickly, which leads to asphyxiation of unaware personnel.

Cryogens are often stored in laboratory dewars, pressurized cylinders, and in large tanks external to buildings. Release of cryogens from laboratory dewars can occur through improper handling procedures. Releases from pres-

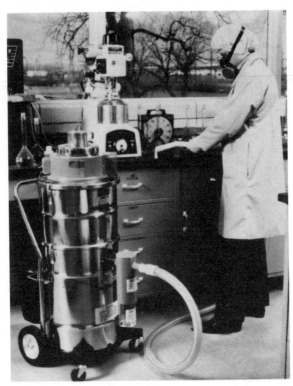

FIGURE 6.3   Special HEPA vacuum cleaner for cleaning up mercury spills. (Used with permission of Hako Minuteman, Inc., Addison, IL.)

surized cylinders can occur through overpressurization, which can occur when the valve and vents become blocked with ice pellets. Releases from tanks can occur when valves, line fittings, or tank walls leak or rupture.

Contact with cryogenic liquids can cause serious burns to eye and skin tissue. Prolonged contact will cause embrittling of materials and flesh. The evaporating gas is cold enough to damage nasal and eye tissues.

The best course of action to deal with a release of cryogenic liquids is to evacuate the area and allow the liquid to evaporate. This may be aided by applying appropriate local exhaust or letting natural ventilation (through open windows and doors) disperse the concentrated gas. Local exhaust might mean venting the gas to a nearby fume hood, exhaust vent, or outside window by a duct and fan. Special care should be used with cryogenic liquids whose gases are flammable (hydrogen) or toxic (carbon monoxide) or support combustion (oxygen). Further information may be found in references 41 and 42 or obtained from the supplier.

6.4.4   *Releases of Compressed Gases*   Compressed gas cylinders are frequently used in laboratories and stored in a chemical storage area. There is a

wide range of compressed gases whose hazardous properties vary substantially in toxicity, flammability, and reactivity. The sudden release of compressed gases in a nonventilated area can create a confined-space effect, causing an oxygen deficiency and asphyxiation of personnel entering the area. Likewise, a concentrated cloud of toxic gas released from a gas cylinder can be fatal in minutes; a cloud of a flammable or reactive gas can result in fires or explosions.

Active measures should be taken to prevent and detect releases of compressed gases. Protecting the valve stem of gas cylinders during transport and handling by keeping the valve cap in place is essential. Inspecting gas cylinders for damage should be part of the receiving process. Dented, bulging, or deteriorating cylinders should be returned to the supplier.

Large cylinders of hazardous gases that are attached to a manifold system should be fitted with an alarm that is activated by a sudden drop in pressure. Similarly, small cylinders of toxic gases should be stored in an area with its own local exhaust ventilation and treatment system. Gas detectors should be installed that can sound an alarm if the airborne concentration surpasses a given level. In storage areas where ventilation does not exist and cylinders of inert gases are stored, both a gas detection alarm and an oxygen deficiency alarm should be in place.

Responding to leaks and releases from compressed gas cylinders will depend on the type of gas, the rate of the release, and the size of container. The response to a leak of nitrogen from a lecture size bottle in a large room is quite different from the escape of hydrogen cyanide. Preplanning the response might include the strategic location of local exhaust ventilation and deployment of scrubbers, collectors, or secondary reaction vessels (Table 6.7). Stringent EPA regulations derived from the Clean Air Act prohibit the discharge of untreated toxic gases to the atmosphere. Additional information may be found in references 41 and 42 or from the compressed gas supplier.

6.4.5 Decontamination

Once a released chemical has been recovered, the area in which the release occurred and any affected equipment or surface must be decontaminated. The decontamination process will vary with the type of chemical, the contaminated surfaces, and the methods necessary to carry out the decontamination. In the cases of acids and bases, decontamination often consists of scrubbing the contaminated surfaces with a neutralizing agent and rinsing with water. Using wetted pH paper to test for neutrality is an effective measure of decontamination. In the case of organic materials, decontamination may be checked with sample wipes via gas chromatography.

Any emergency equipment used in the cleanup of spills should also be decontaminated, especially reusable personal protective equipment like respirators or gloves. The use of disposable clothing where permitted can help limit the amount of decontamination time following a chemical release.

TABLE 6.7   Choices for Treating Releases of Toxic Gases

Gas	Wet Scrubbers[a] Spray	Venturi	Packed	Dry Scrubbing, Adsorption[b]	Burner[c]
Arsine				X	
Chlorine		X	X		
Diborane		X	X	X	
Disilane		X	X		
Fluorine	X		X		
HBr	X	X	X		
HCl	X	X	X		
HF	X	X	X		
H$_2$S		X	X	X	
Phosgene		X	X		
Phosphine				X	X
Silane		X	X	X	X
Silicon tetrachloride		X	X		
Silicon tetrafluoride	X	X	X		

[a] *Wet scrubbers:* Enclosed chambers where a gas is dissolved in liquid to chemically treat the dissolved gas to form a more a stable, les hazardous by-product.

[b] *Adsorption:* A form of dry scrubbing, in which gas molecules adhere to the surface of an adsorbing material, such as adsorption. Chemisorption, another form of dry scrubbing, involves a chemical reaction that changes the gas into a new substance.

[c] *Burning:* The oxidation of a gas in a flame through an incineration process.

Source: Used with permission of Matheson Gas Products, Secaucus, NJ.

Subparagraph (k) of 29 CFR 1910.120 specifies decontamination procedures pertaining to hazardous substance exposure. The standard requires the development and communication of a decontamination procedure, which includes (a) standard operating procedures to minimize employee contact with hazardous materials; (b) decontamination of all employees; (c) monitoring of the decontamination process by a site safety and health supervisor; (d) a specified location that minimizes further contamination; (e) decontamination or disposal of emergency equipment, supplies or solvents; and (f) decontamination of personal protective clothing and equipment. The standard also prescribes requirements for laundries, showers, and change rooms, where appropriate.

6.4.6   Emergency Equipment Response Centers

Once the appropriate cleanup equipment has been chosen, a readily accessible area should be designated as an emergency response center. In this area, the necessary absorbents, neutralizers, protective gear, and other paraphernalia should be assembled, inventoried, and stored until use. It is important to establish a regular inventory—inspection system so that depleted stocks may be replenished to have sufficient supplies on hand for future spills. Having a mobile cart or truck available to promptly transfer the necessary materials to the spill scene will speed up the response to the chemical release. One manufacturer has designed such a response unit for small spills (Figure 6.4).

Personnel responding to a chemical release will need access to containment and cleanup procedures, instructions, and chemical hazard information. The emergency equipment response center should have the appropriate data files (described above) available to the response team.

6.5   PERSONAL PROTECTIVE EQUIPMENT

Since the health or lives of the emergency response team is endangered by a chemical release, proper selection and use of personal protective equipment (PPE) is an absolute requirement. The routes of exposure to hazardous chemicals include inhalation and eye and skin contact. Specific types of

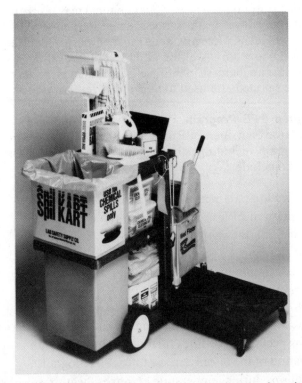

FIGURE 6.4   A mobile cart containing spill cleanup supplies for minor laboratory spills. (Used with permission of Lab Safety Supply Company, Janesville, WI.)

respiratory protection, clothing, gloves, and face and eye protection must be determined in advance for each class and uniquely hazardous chemicals. The emergency response team must be trained in the proper use of PPE. Finally, a designated staff member must inspect, maintain, and repair or restock the protective equipment to ensure its availability and reliability for the next emergency.

6.5.1 OSHA Requirements for Personal Protective Equipment

There are a number of requirements for PPE covered by OSHA's emergency response standard. The standard requires that PPE must be selected and used to protect employees from the hazards and potential hazards likely to be encountered in emergency operations. This selection of PPE must be based on an evaluation of how that equipment will perform relative to the demands and limitations of the emergency response operations.

For conditions where chemical exposure levels present will create a substantial possibility of immediate death, immediate serious illness, or injury, a positive pressure breathing apparatus or positive pressure airline respirator with escape air supply must be worn.

Where skin absorption of a hazardous chemical may result in a substantial possibility of immediate death, immediate serious illness, or injury, total encapsulation chemical protective suits must be worn. Details for these suits are described in Appendix A of the OSHA standard. If the hazards at the release site call for an increase in the level of personal protection, appropriate PPE should be selected and used to afford that level of protection.

6.5.1.1 Written PPE Program For emergency response operations, an employer must have a written PPE program. Specific elements must comprise the written program, including policy statements, procedures, and guidelines. These elements include:

1. PPE selection parameters based upon site hazards
2. Description of PPE use and limitations
3. Work mission duration of specific PPE
4. PPE maintenance and storage
5. PPE decontamination and disposal
6. PPE training and proper fitting
7. PPE donning and doffing procedures (manufacturer directions may be attached to the plan without being rewritten into the plan)
8. PPE inspection procedures (before, during, and after use)
9. Evaluation of the effectiveness of the PPE program
10. Limitations during temperature extremes, heat stress, or other appropriate medical considerations

Two objectives of the PPE program should be to protect the wearer from safety and health hazards and to prevent injury to the wearer from the incorrect use or malfunction of the equipment. Copies of the written PPE program should be made available to all affected employees and a reference copy should be maintained at the release site. Other essential technical information, maintenance manuals, and so on should be gathered, organized, and maintained for reference.

6.5.1.2 *Levels of Protection*   Appendix B of the standard specifies four levels of protection with accompanying lists of PPE depending on the protection required. These levels are described below, and PPE requirements are depicted in Table 6.8.

Level A Protection   Used when chemicals are identified or suspected to be present that require the highest level of protection for skin, eye, and respiratory system. Also includes confined, poorly ventilated areas.

Level B Protection   Used when chemicals are identified or suspected to be present that require the highest level of protection for the respiratory system but do not present a severe skin hazard. Level B also includes atmospheres with less than 19.5% oxygen levels.

Level C Protection   Used when chemicals or their atmospheric contaminants, liquid splashes, or other direct contact will not affect exposed skin, and the type of air contaminants have been identified and measured, and an air purifying respirator can effectively remove the contaminants.

Level D Protection   Used when the atmosphere contains no known hazard, and emergency operations do not anticipate splashes, immersion, or the potential for unexpected inhalation of or contact with hazardous levels of any chemicals.

6.5.2 Respiratory Protection

Inhalation of toxic gases, mists, vapors, particulates, or any combinations thereof are possibilities when dealing with chemical spills. Respiratory protection should be selected based on the level of protection required by the situation (see Table 6.8).

   A requirement for respiratory protection is that all employees, while engaged in response efforts to releases of hazardous chemicals that present an inhalation hazard, must wear a positive-pressure self-contained breathing apparatus (SCBA). The Incident Commander, after monitoring the air, may downgrade to a different type of respirator.

TABLE 6.8 Personal Protective Equipment Required for Emergency Response Operations[a]

Type of Protection	Level A	Level B	Level C	Level D
Respiratory Protection				
Positive pressure SCBA or	X	X		
Positive pressure airline with escape SCBA	X	X		
Air purifying respirator			X	
Escape mask			O	O
Protective Clothing				
Total encapsulating chemical suit	X			
Disposable protective suit	X			
Disposable protective suit	X			
Chemical-resistant clothing, hooded		X	X	
Coveralls	O	O	O	X
Long underwear	O			
Hand Protection				
Outer gloves, chemical-resistant	X	X	X	O
Inner gloves, chemical-resistant	X	X	X	O
Other				O
Foot Protection				
Boots, steel shank/toe (chemical-resistant)	X	X	X	X
Boot covers, disposable (chemical-resistant)	X	O	O	O
Head/Eye/Face Protection				
Hardhat	O	O	O	O
Faceshield		O	O	O
Safety glasses/chemical splash goggles				X

[a] X, Required as specified in Appendix B, 29 CFR 1910.120; O, optional, depending on conditions, as specified in Appendix B.

Spills of harmless solids may produce dusts that require only a NIOSH-certified disposable dust mask. Spills of toxic solid chemicals require greater protection. A NIOSH-certified HEPA filtered cartridge respirator or gas mask will give protection for those toxic dusts for which the respirator manufacturers recommend usage. Some dusts (e.g., carcinogens and cyanides) pose chronic hazards or are lethal in minute quantities. Using a NIOSH-certified SCBA may be the only safe alternative for dealing with toxic dusts.

Liquid chemical releases are far more complex when determining the proper respiratory protection. Liquids can emit organic vapors, acid gases, mists, or combinations thereof that may or may not be effectively stopped by an air-purifying respirator. The concentration and hazardous nature of these vapors or gases is critical for selecting the correct respirator. The vapor concentration illustrated in Tables 6.1 and 6.2 can approach or exceed the concentration level determined to be immediately dangerous to life or health. With this in mind, the scene of a liquid release should best be considered an emergency that requires the use of a NIOSH-certified SCBA.

Releases of compressed gases and cryogenic liquids are dangerous when entering the affected area. The released gas may quickly reach toxic concentrations or if nontoxic create an oxygen deficiency similar to that in a confined space. Air-purifying respirators and gas masks should not be used when confronting a release of compressed gases; positive-pressure SCBAs should be worn.

The possession and use of respirators makes an organization liable to implement the minimally acceptable respirator program specified by OSHA (43). This program includes several elements:

Written standard operating procedures for selection and use of respirators

Respirators selected based on the hazards of the air contaminant

Employees instructed on respirator use and limitations

Regular cleaning and disinfection of equipment

Clean and convenient storage

Regular inspection for worn or missing parts

Medical evaluation of respirator wearers

Wearers to check the fit each time respirators are used

Monthly inspection of SCBAs (44)

Personnel must be fitted with a given respirator, tested for proper fit, and trained in the use of each respirator they are to wear. This fitting and training must occur before personnel respond to an emergency. Training should include a combination of instruction and hands-on practice. Manufacturers' instructions, videotapes, computer-based training, and training booklets can help personnel gain conceptual knowledge of various respirators. A valuable guide to the operation of an SCBA is *Self-Contained Breathing Apparatus, 2d ed.,* by IFSTA (45). Additional helpful information is available for selecting respirators (46,47). Manufacturers of approved and accepted respirators can be found by consulting the NIOSH-certified equipment list (48).

Using a respirator places a strain on the wearer's cardiovascular system and stress on the breathing cycle. Personnel who are selected for the emergency response team must be approved by a physician who has tested and approved their physical capacity to wear and use a specified respirator.

6.5.3 Protective Clothing

Accidental skin contact with chemicals can occur readily when responding to a release of laboratory chemicals. The use of protective clothing to provide a barrier to prevent this contact is essential. The primary trait of effective protective clothing is its ability to hold out vapors, liquid, and particulate matter.

Permeation occurs when liquids in contact with the exterior surface of a material diffuse molecularly through the material so that the chemical appears on the inside. The amount of time for permeation to occur is called the breakthrough time. The rate of permeation may be rapid or slow, so that minute or sizable quantities of a liquid or gas can unexpectedly pass through the material and make contact with skin. There has been a large amount of data on material permeation in the past few years. The breakthrough time and permeation rate depend on the material, its thickness, and the particular chemical.

Penetration of a material by a chemical occurs because of small imperfections or lesions in the material itself. Penetration most commonly occurs at the seams of clothing or gloves.

The level of protection (described in Section 6.5.1.2) required by the situation should guide the selection of protective clothing. Appendix B of OSHA's emergency response standard recommends that protective clothing meet the criteria of the following NFPA standards (49):

Level A   NFPA 1991—Standard on Vapor-Protective Suits for Hazardous Chemical Emergencies

Levels B and C   NFPA 1992—Standard on Liquid Splash-Protective Suits for Hazardous Chemical Emergencies

Level D   NFPA 1993—Standard on Protective Garments for Support Functions at Hazardous Chemical Operations

The NFPA standards cited above call for laboratory performance testing and certification of that performance through garment labeling. For NFPA 1991 and 1992, this performance is validated through a test battery of chemicals, specified in the standards. With each vapor-protective suit and liquid splash-protective suit, manufacturers are required to provide instruction and information on cleaning instructions, marking and storage instructions, frequency and details of inspections, maintenance criteria, how to use test equipment (where applicable), methods of repair (if recommended by the manufacturer), and warranty information.

In addition, manufacturers must furnish training materials that address, at a minimum, donning and doffing procedures, safety considerations, storage conditions, recommended storage life, decontamination recommendations and considerations, retirement considerations, closure lubricants (if appropriate), and visor/faceshield antifog agents or procedures.

NFPA 1991 requires manufacturers to supply chemical permeation resistance documentation; NFPA 1992 requires manufacturers to provide chemical penetration resistance documentation.

Aprons, disposable coveralls, or totally encapsulated chemical splash suits may be demanded by the size of the release, the toxic, corrosive, or hazardous nature of the chemical, and the demand of the emergency response situation. Certified disposable clothing that is impenetrable to liquid and particulates may fit the needs of many laboratory chemical releases. Disposable jumpsuits or coveralls have the advantage of being lightweight, easily stored, and relatively inexpensive. Features such as elastic wristbands, an attached hood, and boots help to provide comprehensive covering for the body.

6.5.4 Gloves

Gloves protect the hands from exposure to chemicals. In many emergency operations the hands of response personnel are prone to come in contact with a chemical. Skin contact with corrosives can cause chemical burns, irritation, or dermatitis. Some chemicals may penetrate or permeate the skin and become absorbed in the bloodstream in small yet toxic doses. The proper selection of protective gloves to prevent skin contact with the chemical is of prime importance. Gloves must be selected on the basis of resistance to (a) chemical degradation, (b) permeation, and (c) penetration.

There is no glove that is free from degradation by all chemicals. Contact with certain chemicals will cause softening, swelling, embrittling, cracking, and dissolving of the glove material so that the barrier properties of the glove are lessened, increasing the chances of skin contact. Consulting the glove manufacturer's literature and even field-testing glove materials will guide the selection of the proper glove for use with a specific group of chemicals. Choosing an appropriate glove for protection from inorganic acids and caustics is much easier than choosing a standard glove for a variety of organic solvents. Many glove materials are made of plastic or rubber. Contact with organic solvents will cause different degrees of degradation to those glove materials. For the best preparation, a facility may need to stock several pairs of gloves made from different materials. The response team can then choose the appropriate one to effectively resist degradation from the spilled organic liquid.

A convenient compilation of glove and protective clothing performance is provided on the GlovES+ microcomputer software and its accompanying *Chemical Protective Clothing Performance Index* (50). A user may query the database by chemical or chemical group and compare the effectiveness of different types of materials against that chemical.

Glove manufacturers like Edmont Wilson, Pioneer, and others provide a guide to chemical permeation for their glove materials and a list of common chemicals. Viton (51) and butyl rubber gloves offer varying resistance to permeation and high resistance to chemical degradation. However, the reader should consult manufacturers' literature when evaluating which gloves to inventory and designate for chemical spills. Another option for anticipating the performance of glove and protective clothing materials is to consult with independent testing laboratories, such as Texas Research Institute (52), which provide objective analysis and certification of protective clothing.

In cases where critical protection of the skin tissue is necessary and a seamed glove is the glove of choice, the practice of submerging an inflated glove in water to detect leaks presents another option. Commercially available glove inflators, such as those used for electrical utility lineworker's gloves, might be considered for this type of test.

6.5.5 Eye/Face Protection

American National Standards Institute Standard Z87.1-1989 (53) requires that chemical splash goggles be used whenever there is a danger of chemical splash injuring the eyes. Industrial hygienists and physicians recommend that chemical contaminants in eyes be flushed with water for 15 minutes. Preventing eye contact with chemicals is essential. Chemical splash goggles have indirect or baffled ventilation ports that direct any liquid entering the goggle away from the eyes. In some instances face shields may be necessary to protect other parts of the face from chemical splash. For chemicals that are lacrymators (eye irritants), a full facepiece respirator that seals against the face may be required.

The chemical release scenario should be imagined creatively to anticipate worst case and realistic possibilities. This exercise can help identify the various needs for protective equipment, which can be ordered and inventoried before the actual need arises.

6.6 THE EMERGENCY RESPONSE TEAM

The emergency response team is the crux of the entire effort to clean up chemical releases efficiently and safely. Response personnel must be able to meet stringent criteria and be on call to respond with little or no notice. Ability, availability, and reliability are key attributes of response staff. Simply selecting personnel based on a positive work attendance record is not enough. Criteria for a successful spill response team include personality and psychological factors and medical and physical fitness.

Personality traits of prudence, caution, and the ability to deal with the stress of a potentially dangerous situation must be weighed carefully. Since a team effort is required in cleanup situations, team members must exhibit comradery and the ability to work with others. Clear thinking under pressure, the ability to follow instructions, and persistence are necessary.

Physically, members of the emergency response team should be strong enough to use the designated tools, supplies, and PPE, such as respirators. Members of the response team should not be overly sensitive to chemical exposures. Screening candidates' medical history is essential. Pregnant women may be especially vulnerable to chemical exposure and should not be an on-the-scene member of the response team. Conducting the final selection of emergency response candidates should be done through a careful decision matrix.

The full complement of functions that make up the emergency response team is described below. An organization that has organized its team may, because of limited resources, assign multiple functions to its team members. At the minimum, however, no less than two persons (using the buddy system) should comprise an actual response unit. Since dangerous conditions can exist, communication devices (if required by the circumstances) should be used between the cleanup squad and the incident command post.

The most senior facility official responding to the release is the *Incident Commander* who has the minimum following duties: (a) to assess the situation; (b) to activate, brief, coordinate, and control the Emergency Response Team and its operations; (c) to approve and implement an "incident action plan"; (d) to approve requests for resources; (e) to approve decontamination and disposal plans; and (f) to release the facility for resumption of operations. The Incident Commander must have 24 hours of training and certified competency in the knowledge of implementation of the incident command system, the response plans of the organization, the local community, and the state; hazards and risks of employees working in chemical protective clothing; and decontamination procedures.

The *Safety Official*, appointed by the Incident Commander, has a distinctive role in assuring the safety of personnel at the emergency site. The safety official is responsible for identifying and evaluating hazards. In addition, the safety official provides direction with respect to the safety of emergency operations taking place. The safety official has the authority to alter, suspend, or terminate activities when IDLH (immediately dangerous to life or health) or imminent conditions exist. The safety official must immediately inform the Incident Commander what can be done to correct those hazards.

Other personnel involved in emergency response operations include skilled support personnel (from outside the organization), specialist employees, first responders (awareness and operations levels), hazardous materials technicians, and hazardous materials specialists.

Skilled support personnel who assist at the response scene must be briefed on their participation in the emergency response. This briefing should include instruction in wearing PPE, what chemical hazards are involved, and what duties are to be performed.

Specialist employees are those who work with (on a regular basis), and are trained in the hazards of, specific hazardous substances. Specialist employees are called on to provide technical advice or assistance at a hazardous chemical release to the Incident Commander. Specialist employees must receive training or demonstrate competency in the area of their specialty annually.

First responders (awareness level) are individuals who witness or discover a hazardous chemical release and who are trained to initiate the emergency response sequence by notifying the appropriate authorities. These first responders would take no further action beyond notifying authorities. First responders on the awareness level must be aware of what hazardous materials are and the risks present in an emergency release, able to recognize the presence of hazardous materials, able to identify the hazardous chemical, if possible, and aware of their role in the emergency response plan.

First responders (operations level) are individuals who respond to releases or potential releases of hazardous chemicals in a defensive mode. Their role is to contain the release from a safe distance, keep it from spreading, and prevent exposures. First responders are not mean to actually stop the release if it is still in progress. First responders are to have at least eight hours of training, sufficient to be competent in a variety of operations associated with first response at the operations level [see 1910.120 (g)(6)(ii)].

Hazardous materials technicians are individuals who respond aggressively to emergency releases or potential releases to stop an ongoing release. Their mission is "to approach the point of release in order to plug, patch or otherwise stop the release." Ruptures of drums and flows from other large containers that are spouting leaks would be handled by an individual with skills and training suitable to a hazardous materials technician. Hazardous materials technicians must have at least 24 hours of training, with certified competencies for performing the duties assigned to them as HAZMAT technicians [see 1910.120 (q)(6)(iii)].

Hazardous materials specialists are individuals who respond with and provide support to hazardous material specialists. In addition to performing duties similar to technicians, specialists' duties require a more specific knowledge of hazards and substances. Hazardous materials technicians must have at least 24 hours of training, with certified competencies for performing the duties assigned to them as HAZMAT technicians [see 1910.120(q)(6)(iv)].

Training for all responders must be conducted before they participate in actual emergency operations. Trainers must meet OSHA criteria for instruction and have knowledge of the subject matter. Refresher training must be provided annually to maintain competency. In lieu of such training, employees may demonstrate their competency, which must be documented by their employer.

Training of the team must include familiarization with the types and hazards of laboratory chemicals, emergency response methods, the designated chemical hazard information, and protective equipment. Hands-on experience during drills and simulated releases must be provided. Safety precautions and disposal methods should be reviewed periodically. Cleanup equipment and its use must be understood prior to the emergency. Training resources and activities might include lectures, films and videotapes, computer-based and interactive video training, and demonstrations. The application of first aid procedures should be studied and reviewed. Assessment of training programs of other organizations can cause a beneficial information exchange to help broaden the training base (54–57). More detail may be found in 29 CFR 1910.120(q).

6.7 CONCLUSION

Releases of laboratory chemicals may be large or small, hazardous or harmless. Many releases can be prevented by using simple management and engineering controls, such as inspecting container conditions, instituting safe handling procedures, and providing a compatible and secure storage environment. For releases that do occur, an informed plan of action, properly selected equipment, and a well-trained response team can turn a potential disaster into a routine matter. The safe storage of laboratory chemicals is enhanced by properly planned and managed emergency response tactics.

REFERENCES

1. J. P. Dux and R. F. Stalzer (1988). *Managing Safety in the Chemical Laboratory,* Van Nostrand-Reinhold, New York, p. 31.
2. "Hazardous Waste Operations and Emergency Response," *Code of Federal Regulations,* Title 29, Part 1910.120, 1989.
3. D. Hedberg. "Spill Control in the Chemical Storeroom," presented at the *187th National ACS Meeting,* Kansas City, September 1982.
4. W. T. Niggel, "Leakage and Overpressurization of 55-Gallon Drums," in the *Proceedings of the 1982 Hazardous Materials Spills Conference,* Milwaukee, WI, April 1982, pp. 482–486.
5. National Research Council (1988). *Prudent Practices for Handling Chemicals in Laboratories,* National Academy Press, Washington, DC.
6. M. Lee. "Development of a Local Hazardous Materials Management System" in the *Proceedings of the 1982 Hazardous Materials Spills Conference,* Milwaukee, WI, April 1982, pp. 147–150.
7. D. Austin. "Spill teams, industrial responders and the Incident Command System" *Hazmat World,* April 1989, Vol. 2, No. 4, pp. 16–19.
8. U.S. Coast Guard (1986). *CHRIS Hazardous Chemical Data,* Government Printing Office, Commandant Instruction M16465.12, October, 1978 (also available from Noyes Data Corporation as *Hazardous Chemicals Data Book,* 2d ed., G. Weiss (Ed.), Park Ridge, NJ 1986).
9. *Handling Chemicals Safety,* Dutch Association of Safety Experts/Dutch Chemical Industry Association/Dutch Safety Institute, Netherlands, 1980.
10. *Toxic and Hazardous Industrial Chemicals Safety Manual,* International Technical Information Institute, Tokyo, Japan, 1988.
11. *NIOSH-OSHA Pocket Guide to Chemical Hazards,* U.S. Government Printing Office, Washington, DC, 1987.

12. N. Sax and R. J. Lewis (1988). *Dangerous Properties of Industrial Materials, 7th ed.*, Von Nostrand-Reinhold, New York.

13. *Emergency Response Guidebook for Hazardous Materials, 1987*, U.S. Department of Transportation (printed by Labelmaster, Chicago, 1987).

14. General Electric Company (1980). *Material Safety Data Sheets*, Schenectady, NY.

15. *Chemical Information Manual*, Government Institutes, Inc., Baltimore, MD, 1988.

16. N. Proctor et al. (Eds.) (1989). *Chemical Hazards of the Workplace, 2d ed.*

17. N. I. Sax and R. J. Lewis, Sr. (1987). *Hazardous Chemicals Desk Reference*, Van Nostrand-Reinhold, New York.

18. M. Sittig (Ed.) (1985). *Handbook of Toxic and Hazardous Chemicals and Carcinogens, 2d ed.*, Noyes Data, Park Ridge, NJ.

19. E. Z. Joseph (Ed.) (1985). *Chemical Safety Data Guide*, Bureau of National Affairs, Washington, DC.

20. D. J. DeRenzo (Ed.) (1986). *Solvents Safety Handbook*, Noyes Press, Park Ridge, NJ.

21. L. Bretherick, (1990). *Handbook of Reactive Chemical Hazards, 4th ed.*, Butterworths, London.

22. C. Foden, *Hazardous Materials Emergency Action Data*, 1989 (available as a subscription service).

23. U.S. EPA, (Jan. 1979). *Hazardous Material Spill Monitoring: Safety Handbook and Chemical Hazard Guide, Part A*, National Technical Information Service Publication No. PB 295853.

24. M. Lefevre (1989). *First Aid for Chemical Accidents, 2d ed.*, Academic, New York.

25. Code of Federal Regulations, Title 40, Sections 261–265.

26. "Computer Systems for Chemical Emergency Planning," *Chemical Emergency Preparedness and Prevention, Technical Assistance Bulletin 5*, U.S. EPA, Office of Solid Waste and Emergency Response, OSWER 89-005, September 1989.

27. American Conference of Governmental Industrial Hygienists (ACGIH) (1989). *TLV's—Threshold Limit Values for Chemical Substances and Physical Agents in the Workplace*, ACGIH, Cincinnati, OH.

28. These must be calculated. For a discussion of evaporation rates, see the article by D. Machay et al., "Calculation of the Evaporation Rate of Volatile Liquids" in *Proceedings of 1980 National Conference on Control of Hazardous Material Spills*, Louisville, KY, May 1980.

29. D. Pipitone and D. Hedberg (May 1982). "Safe Chemical Storage: A Pound of Prevention is Worth a Ton of Trouble," *J. Chem. Ed..*

30. Reference 5.

31. J. Meidl (1978). *Flammable Hazardous Materials*, 2d ed., Glencoe Press, Encino, CA, p. 168.

32. Reference 10.

33. Z. R. Luce (Oct. 1989). "A Safety Program for an Explosives Manufacturing Environment," *Professional Safety* **34**(10), 35–38.

34. Hazardous Material Response Kits, which include devices for plugging drums and other containers are available from Edwards and Cromwell Mfg, Baton Rouge, LA.

35. M. Pitt (Oct. 16, 1982). "A Vapour Hazard Index for Volatile Chemicals," *Chem. Ind.*, 804–806.

36. D. Hedberg (Mar. 1981). "Clean Up of Chemical Spills in Labs," *Natl. Safety News*.

37. Reference 10.

38. J. Brugger, "Selection, Effectiveness, Handling and Regeneration of Sorbents in the Clean Up of Hazardous Material Spills," in *Proceedings of 1980 National Conference on Control of Hazardous Material Spills,* Louisville, KY, May 1980.

39. Federal Register, Monday, May 19, 1980, Code of Federal Regulations, Title 40, Section 261.33, Paragraphs (e) and (f).

40. Code of Federal Regulations, Title 40, Section 261.5(c)(4).

41. *Guide to Safe Handling of Compressed Gases*, Matheson Gas Products, Inc. Secaucus, NJ, 1982.

42. Compressed Gas Association (1982). *Handbook of Compressed Gases*, 2d ed., Van Nostrand-Reinhold, New York.

43. Code of Federal Regulations, Title 29, Section 1910.134.

44. J. F. Rekus (Oct. 1989). "There's More to Respirators Than Meets the Eye," *Occupational Health and Safety*, 50–59.

45. IFSTA (1990). *Self-Contained Breathing Apparatus*, 2d ed., Fire Protection Publications, Oklahoma State University, Stillwater.

46. ANSI Standard Z88.2-1989 (1989). "Practices for Respiratory Protection," American National Standards Institute, New York.

47. J. Pritchard (1976). *A Guide to Industrial Respiratory Protection*, DHEW Publication No. 76-189, NIOSH, Cincinnati, OH.

48. *NIOSH Certified Equipment List*, DHHS Publication No. 81-144, NIOSH, Cincinnati, OH.

49. NFPA Standards 1991, 1992, 1993 are published by the National Fire Protection Association, Quincy, MA, 1990.

50. K. Forsberg, *Chemical Protective Clothing Performance Index with GlovES+*, Instant Reference Sources, Inc., Austin, TX.

51. Viton is a registered trademark of the DuPont Company.

52. Texas Research Institute, Inc., Austin, TX.

53. ANSI Standard Z87.1-1989 (1989). "Practice for Occupational and Educational Eye and Face Protection," American National Standards Institute, New York.

54. G. Tompkins and R. Garton, "Hands On Training for Industrial Emergency Response Teams," in *Proceedings—1982 Hazardous Materials Spill Conference*, Milwaukee, April 1982, pp. 497–499.

55. J. F. Rekus (Oct. 1989). "Disaster Drills Identify Potential for Problems in Real Emergencies," *Occupational Health and Safety*, 126–130.

56. J. T. Baker Inc., Hazardous Chemical Spill Response Workshop.

57. *Hazardous Materials Training Handbook*, Bureau of Labor Relations, Washington, DC.

CHAPTER 7

COMPUTER APPLICATIONS AND SYSTEMS FOR HAZARDOUS MATERIALS AND WASTE COMPLIANCE IN THE ERA OF REGULATION

ALLEN G. MACENSKI
Operations Health and Safety Manager
TRW, Inc.
Redondo Beach, California

7.1 INTRODUCTION: THE IMPACT OF REGULATION ON CHEMICAL SAFETY MANAGEMENT

Since the 1970s, Congress has enacted a series of new environmental health and safety laws. These laws included the Occupational Safety and Health Act of 1970, the Environmental Protection Act of 1970, the Resource Conservation and Recovery Act of 1976, and the Federal Water Pollution Control Act. Political scientists have classified these laws as "command and control" laws, that is, laws that set standards for chemicals and control their use and emissions. Federal and state legislation of this nature have dominated the environmental and occupational health and safety scene.

When the Occupational Safety and Health Administration published its final standard for hazard communication for chemical manufacturers on November 25, 1983 (48 FR 53280), it was the culmination of nearly 10 years of federal rulemaking activity by the agency. It was later expanded to include other industries as defined in the *Federal Register* of August 24, 1987. The Hazard Communication Standard requires manufacturers, distributors, users, and importers to provide information to employees on all hazardous chemicals used in the workplace through the use of labels, Material Safety Data Sheets (MSDSs), and training programs. Now the Hazard Communication Standard and the 1986 Superfund Amendment and Reauthorization Act, Title III regulations require information management, handling, and dissemination. This set of requirements has led to some information handling problems.

Under the Hazard Communication Standard, the MSDSs that are to accompany each order or shipment of hazardous chemicals are the primary vehicle for transmitting the detailed safety, industrial hygiene, and risk management information. Hypothetically, both labels and end-user training are to be based on the information contained in the MSDS. To be effective, the MSDS must be complete, accurate, and most of all, up-to-date. Updating is required under the Hazard Communication Standard but also becomes practically important for in-plant and community response purposes. Such updates often include corrections and current research applications.

Both Congress and many state legislative houses, like that of California, have in the wake of chemical release incidents at Bhophal, India, and elsewhere passed legislation providing for community emergency response planning and notification and for information about chemical substances to be made available to the public. In 1986 Congress passed the Superfund Amendment and Reauthorization Act (SARA) which included Title III. Title III provides for emergency response planning. It requires designated state commissions to establish procedures for receiving and processing public requests for information collected under the requirements. In addition, states must coordinate activities of local emergency planning committees and review local chemical emergency plans. This has led to citizens evaluating chemical safety data, defining levels of concern for various chemicals, and developing community-based standards for each chemical.

SARA requires facilities to immediately notify the local emergency planning committee and state emergency response commissions if there is a release of certain extremely hazardous substances. Notification is required for any of the approximately 385 extremely hazardous substances listed by the Environmental Protection Agency (EPA), or any substance subject to emergency notification requirements under the Comprehensive Environmental Response Act/Compensation and Liability Act (CERCLA), Section 103(a), if the release is in excess of the reportable quantity established for the substance.

Many manufacturing facilities in Standard Industrial Classification codes 20–39 are required to submit emissions inventories to inform the EPA, state officials, and the public about releases of certain toxic chemicals (Section 313). For purposes of this toxic chemical release reporting, the EPA has designated substances subject to reporting under New Jersey and Maryland laws, with the option of adding or modifying the list (originally 329 chemicals). This list focuses on substances that meet several criteria: (a) those that are known to cause cancer or serious reproductive disorders, genetic mutations, or other chronic health effects; (b) those that can cause significant adverse acute health effects outside the facility resulting from continuous or recurring releases; and (c) those that can cause adverse effects on the environment because of toxicity, persistence, or a tendency to accumulate.

The emissions reporting requirements apply to those facilities with 10 or more full-time employees in SIC codes 20–39 that manufactured, processed, or otherwise used these chemicals in excess of certain thresholds (use of 10000 pounds or more in any calendar year; manufacture or processing of 75000 pounds or more in 1987 (reported by July 1, 1988 and 25000 pounds or more reported in July 1989 and thereafter).

The sections of Title III most dependent upon the MSDS are Sections 311 and 312, the community right-to-know reporting requirements. Section 311 calls for any facility that is required to prepare or maintain an MSDS under the OSHA Hazard Communication Standard to submit either copies of their MSDSs or lists of the MSDS chemicals to the state emergency response commission, local emergency planning committee, and local fire department. If choosing submission of the list, the chemicals must be organized in categories of health and physical hazards. The state allows the EPA to establish threshold quantities under which facilities will not have to report.

Section 312 requires submission of an emergency and hazardous chemical inventory form to the same officials, in one of two distinct fashions. Tier I involves aggregate information about the estimates of average daily and annual quantities, along with the general location of the chemicals, Tier II along with the general location of the chemical. Tier III information may be requested later, and, among other things, will include the chemical name or common name that appears on the MSDS, estimates of maximum amounts of the chemical present at any given time, location in the facility, description of storage method, and indication of any trade secret claims. Both Tier I and Tier II information must be made available to the public upon request, subject to protection of trade secrets (Section 322).

When Title III was initially enacted, these two sections applied to the manufacturing sector only. However, with the expansion of the scope of the Hazard Communication Standard, many more facilities should be maintaining two sections. There are many unanswered questions in the regulatory impact of Title III, especially for small business.

However, today government tracks the life cycle of a chemical more than ever before (Figure 7.1). And it is anticipated that more regulation will occur, thus creating issues to be resolved in business and community agencies.

7.2 COMPUTERIZED SOLUTIONS FOR CHEMICAL SAFETY INFORMATION REQUIREMENTS

To comply with a diverse set of requirements, many government agencies and companies are seeking innovative management tools to meet the requirements of federal and state standards; others are seeking sources of information on hazards in the workplace; still others want to standardize their MSDSs for ease of retrieval and training efficiency. Many are seeking to integrate their sample records, training records, analytical records, medical records, and

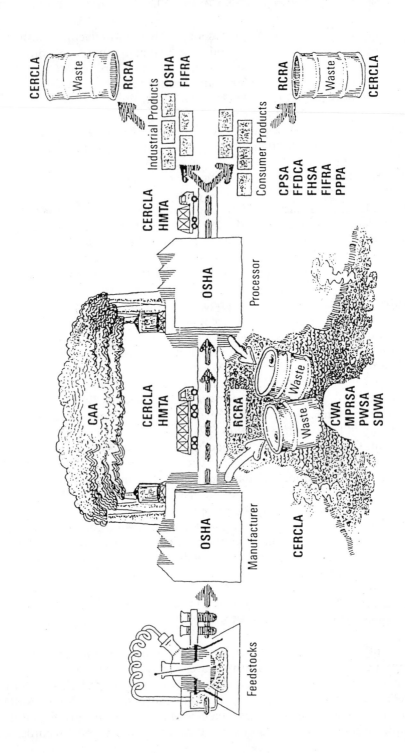

TSCA ⟶ **TSCA** ⟶ **TSCA**

CERCLA	—	Comprehensive Environmental Response Act/Compensation & Liability Act
HMTA	—	Hazardous Materials Transportation Act
OSHA	—	Occupational Safety & Health Act
RCRA	—	Resource Conservation & Recovery Act
TSCA	—	Toxics Substances Control Act
CPSA	—	Consumer Product Safety Act
FFDCA	—	Federal Food Drug & Cosmetic Act
FHSA	—	Federal Hazardous Substances Act
FIFRA	—	Federal Insecticide Fungicide & Rodenticide
PPPA	—	Poison Prevention Packaging Act
CAA	—	Clean Air Act
CWA	—	Clean Water Act
MPRSA	—	Marine Protection Research & Sanctuaries Act
PWSA	—	Ports & Waterways Safety Act
SDWA	—	Safe Drinking Water Act

FIGURE 7.1 How government tracks the life cycle of a chemical.

OSHA compliance records with their MSDSs. This has resulted in a tremendous demand for off-the-shelf labels, MSDSs, and training programs. The computer is now becoming a required tool for the efficient mastery of chemical safety information.

It has often been said that one must crawl before one can walk. Manual reorganization or, in some cases, organization of chemical inventory and MSDS information is often necessary before the software needs and characteristics for a chemical safety data management computer system can be adequately analyzed. The same concepts can be applied to broader computerization of chemical data, as it pertains to total hazardous waste data management (i.e., any program that requires tracking many hazardous chemicals).

Options that are currently available for chemical safety data management include:

1. Manual recordkeeping; manual achieving; manual production, computation, and statistical synthesis of reporting; and manual retrieval of reports
2. Microcomputer plus an off-the-shelf data manager
3. Microcomputer plus a full custom data manager, including support and special technical features
4. Minicomputer plus terminals plus printer
5. Mainframe plus terminals plus printer

Most of-the-shelf data manager MSDS collections are composed of a number of chemical data sheets that are collected from manufacturers and/or enhanced with additional information found in the scientific literature. Some commercial computer programs offer supplemental generic MSDS databases that can be used in addition to or in combination with MSDS information provided by manufacturers or distributors. These databases are built from the literature, from other MSDS-like databases, or from the originating company's own MSDS collection. Some are unique.

7.2.1   Computer Implementation

The optimum information management system will help in obtaining the chemical data (inventory and storage recordkeeping), interpreting the MSDS (self-training and recording of training efforts), distributing MSDSs (tracking and recording the distribution), and tracking waste chemical information.

To meet these requirements, many moderate to large organizations have turned to their in-house mainframe and minicomputers. Automated recordkeeping allows for efficient standardization, fast retrieval, easy update, and low-cost storage. The computer can be used for emergency response (air dispersion modeling), for highlighting safety and health trends, to generate legal reports, and to track MSDS distribution data. The computer can record chemical inventories; organize, file, and update MSDSs; track employee medical and exposure histories; and be used for training programs. Hazardous materials management and regulatory compliance involve identifying and inventorying hazardous substances; obtaining, updating, distributing, and interpreting MSDSs; and tracking, reporting, and manipulating sampling data, personnel medical records, incident reports, and other health and safety data.

Several off-the-shelf integrated occupational health and safety information systems with chemical safety modules are available for these organizations. Although systems have been successfully developed with user companies by in-house information systems groups, it is almost always less expensive and time-consuming to purchase or time-share flexible, commercially available systems with customization and maintenance by in-house computer specialists. The microcomputer offers some advantages that cannot be achieved easily by the giant mainframe. The most important is its availability. An additional advantage is that there is security for sensitive information. Access to a microcomputer can be restricted to those people with a definite need to know. In addition, access within microcomputer programs also can be controlled.

The advantages of automated health and safety systems can now be enjoyed by smaller companies and small health and safety offices with the advent of microcomputers and the recent proliferation of specialized software to handle such data as MSDSs and other environmental and occupational health information needs. The microcomputer offers the advantages of availability of data retrieval and security access.

7.2.2   CD-ROM: Information Technology Using Microcomputers

Advances in data architecture technology now make it possible to locate precise information without spending hours, or days, searching files previously stored on hard or soft disk systems.

Today, volumes of data can be stored digitally and accessed instantly with the help of laser technology and microcomputers. The most common form of this electronic media is Compact Disk Read Only Memory (CD-ROM). With CD-ROM, the microcomputer is provided the power, convenience, and versatility once found only with on-line computer searching. In addition, CD-ROM does not require a librarian or professional searcher to act as an intermediary. CD-ROM's ease of use allows end users to manipulate their own search to find information quickly. This is of special note to health and safety professionals who require fast access to vast data fields to assist in critical decision making.

CD-ROM avoids the data transmission rate constraints, costs associated with the telecommunications links, and problems with communication protocols. It is well suited for countries where phone links are costly and unreliable. In addition, CD-ROM can deliver extensive on-disk help facilities and user tutorials, which make periodic searching more user-friendly, an invaluable asset to the novice.

Many subjects are published on CD-ROM, such as occupational health and safety, industrial hygiene, toxicology, epidemiology, ergonomics, standards, and programmed education and training.

There are CD-ROM databases devoted to specific chemical databanks which describe chemicals by a range of standard data elements: physical, chemical, environmental, health, regulatory, and other characteristics. Additionally, CD-ROM databases offer general information on medicine, environmental impact, hazardous waste, pollution, emergency response, hazardous materials packaging, and transportation regulations.

7.3   IMPLEMENTATION CONSIDERATIONS

Computerization of vital chemical safety records forces a standardization in the long term, but in the short term your next question may be, "What use will I get out of computerization?" Initial data entry and software may be costly, but the long-term benefits must be defined and justify these costs. For most businesses, under the present regulatory climate, the answers to the following questions must be evaluated carefully:

1. Do you spend more than two hours a day filling out forms in duplicate?
2. Would additional statistics or data assist in controlling losses, inventory, or preventive actions to be taken?
3. Do vital records often get misplaced?
4. Is the delay in receiving vital information from distant corporate locations becoming excessive?
5. Are you required to supply file information over the telephone, and do you often respond with the dreaded answer, "I'll call you back when I find it in the files?"
6. Could you use another clerical worker to assist with the paperwork?
7. Are you becoming concerned about regulatory reporting requirements from local, state, or federal agencies?

If your answer is yes to more than one of these questions, then you or your department can improve its efficiency and loss control efforts with the aid of a business microcomputer and appropriate software.

7.3.1   Software Features

There are some standard features you should look for in all software programs. These include index functions and "help" capabilities; menu-driven features with optional labelmakers, MSDS tracking capabilities, sufficient field length to encompass large texts on chemical health hazards and spill precautions, along with enough fields to accommodate optional environmental data that may be required for your local, state, and federal reporting. The ability to search rapidly and display select parts of your overall inventory or a particular MSDS is also important. This index function enables you to search for a specific MSDS or, in some cases, to view a list of all MSDSs. This is the porthole to the MSDS file and, if the program is well structured, can lead the user to the appropriate MSDS even if just a few letters of the chemical name are used in the search. The optimum program should also be able to search for classes of chemicals such as all halogenated compounds. This generally is done through a string function command or index. With this capability, ad hoc questions are simply to formulate and to generate reports containing the requested information.

On-line "help" functions allow the program user to get help and instruction whenever the appropriate key is depressed. Such functions are essential for the novice and occasional computer user. Having a help function and menu-driven capability can lead the first-time user through the query maze with minimal frustration.

Choosing the right system will also include evaluating response time for a query or print. Depending on user requirements for timeliness, it may take up to 30 seconds for the generation of one MSDS and to begin to print the contents. In an emergency every second counts, but you may want to look at other features before eliminating a program solely based on report generation/print response time, especially when emergency response may not be critical in your anticipation.

Chemical safety software systems generally are unique; very few programs are designed with the same approaches to data architecture. One vendor may have discovered a module such as mass balance form, that no other vendor has thought about. Thus, the user must evaluate systems closely for functionality. Today's software dealing with hazardous materials management generally sees vendors providing such items as facility questionnaires, waste management modules, tracking modules, and quick retrieval tools. Consider all your chemical safety management needs so that you do not have to expand your system or buy separate software to incorporate other needs, such as training, permit tracking, and emergency response data.

If a supplemental MSDS database is provided with the chemical safety management program, the frequency of updating those MSDSs sold is a critical question. Under the OSHA Hazard Communication Standard, new information on a chemical is to be placed on the MSDS within 90 days of its publication. Likewise, for emergency purposes under Title III, one would like the most up-to-date information at all times.

A database that is not updated regularly can subject you to a potential citation for a violation of various chemical safety related laws, as could out-dated MSDSs. These commercially available supplemental databases can overlap, may not give sufficient information for the area of your interest, and may not be offered in widely varying formats.

For some users the availability of the capacity to input SARA Title III, Tier I, and Tier II required data may be significant in their purchase decision. Before rejecting a software program that may lack this capability, remember to ask the vendor about the program's expansion possibilities and any imminent modifications on proposed revisions.

Since the typical health and safety microcomputer is usually used for more than just MSDSs, the MSDS program should be able to be easily and quickly retrieved when necessary and not tie up the entire resources of the microcomputer. Some early programs literally required a dedicated microcomputer and a computer-competent operator. Some functions of even more recent programs require a fair amount of computer expertise. An extra investment in a large hard disk, expanded core memory, CD-ROM, or a new operating system may be required before some microcomputers can run some commercial MSDS data file programs.

In addition to having features that facilitate obtaining and interpreting the MSDS, a module of a software program for MSDSs should have the capability to track the distribution of each MSDS. This could involve distribution to other companies, employees, community groups, and so on. Such a record could prove invaluable in supporting employers' past actions in compliance issues and contested cases of product or employee liability.

7.3.2   Microcomputer MSDS Programs with Generic MSDS Database

For a small company without the resources to immediately input all manufacturers' data sheets, an off-the-shelf microcomputer software program with a supplemental MSDS database may meet a critical need. These generic MSDS programs can also be used to check manufacturers' MSDSs, to add to the available safety and health reference data, and to train employees in the use of hazardous substances. Though off-the-shelf microcomputer programs that will handle MSDS data input and then will format documents abound, the marketplace for those with supplemental MSDS databases is much more limited.

In addition to the initial purchase costs of the software, the long-term investment required to install, maintain, and operate the full working system must be considered. Such costs will include assembling the required data, validating and entering the data, training new personnel, purchasing updated software, and correcting and amending the data as changes occur. These costs will apply to some extent to any software application purchased.

7.4   CONCLUSION

It is the objective of government to safeguard the health and well-being of its citizens, and to protect the environment. In the past five years these goals, coupled with an increasingly vocal electorate, have focused relentless political pressure on legislators to tighten standards for dealing with hazardous materials. Consequently, the number of pages of regulations governing health, safety, and the environment more than doubled during this period.

Companies that use, store, manufacture, and transport hazardous materials are facing increasingly complex reporting requirements. American corporations were not ready to handle the regulatory explosion of the 1980s, and looming in the 1990s are new and stricter environmental laws. The current federal environmental agenda includes the pending Clean Air Acts amendments, the proposed reauthorization of the Resource Conservation and Recovery Act (RCRA), a proposal to amend current pesticide control laws, and proposed changes to the Superfund law which expires in 1991. At the state level, regulatory activity is speeding up to the point that regulation is expected to take on tidal wave proportions. (Consider California's Proposition 65-type legislation.) Figure 7.1 illustrates how the federal government currently regulates chemical commodities and the risk they present to citizens.

Many small and medium-sized firms have been vainly waiting for some of the current laws to "go away." Probusiness regulators have fallen from favor as the public looks for lawmakers that are tough on emissions and cleanups. More emphasis is being placed on enforcement, and corporate directors and officers can now go to jail for failing to comply with chemical safety regulations. Nevertheless, top management in many firms resist change and regard safety and environmental affairs departments as cost burden centers. More than any other factor, the threat of big fines and public notoriety is most likely to continue to promote compliance. In the decade ahead, government regulations are expected to continue to drive the need for software designed for the management of hazardous substances.

Many factors enter in the selection process for chemical safety software and training packages. Each available product must be judged by how well it meets the needs of the purchaser. A program that matches your needs or can be customized to meet your needs is the most important criterion for selection on any off-the-shelf commercial product. Let the buyer beware; in the rapidly growing and competitive marketplace, it is recommended that new users contact experienced users and/or user groups for their assessment and expertise with both the program being offered and the vendor support. And finally, the effective allocation of resources will require the elimination of redundant hardcopy systems, the sharing of data, and an investment in flexible modular systems. Development of shared chemical safety information management systems requires careful planning. The following performance criteria are recommended in evaluating each software package before purchase:

1. Emergency response planning information (e.g., hazardous materials and facilities, locations, characteristics, training)
2. Air dispersion modeling (e.g., releases, gas clouds)
3. Other environmental monitoring (e.g., water, groundwater, chemical properties)
4. Facility environmental monitoring and other chemical and waste data (e.g., monitoring data, schedules)

5. Facility chemical or waste recordkeeping, reporting, and compliance assistance (e.g., manifests, labels, report generation)
6. Treatment/pretreatment assistance (e.g., recordkeeping)
7. Facility or treatment design assistance
8. Cleanup assistance
9. Facility maintenance and equipment monitoring and repair
10. Facility permit applications assistance (e.g., NPDES and RCRA Part B)
11. Facility operations and management assistance (e.g., budget-keeping management records)
12. Chemical and properties reference source (e.g., MSDS information)
13. Regulatory reference data source
14. Federal/state information source (e.g., historical accident and incident recordkeeping)

The vendor should be contacted to determine the extent to which the system addresses specific needs and to verify the system's capabilities.

7.5 APPENDIX

This section contains a list of computer software applications and has been assembled as a reference source to assist in locating potentially useful software applications. The systems included have been identified from readily available information sources. The principal intent is to identify software that is applicable to the information collection, data management, reporting, planning, or scheduling requirements of the various regulatory requirements that affect chemical substances. The following list is not comprehensive, nor does it imply approval for any of the resources listed.

Bibliography and Resources

Asbury, A. J. (Sep. 24, 1983). Computers in medical education *Br. Med. J.* **287**, 887–890.

Barnes, K. (May 29, 1984). Good screen design: An important asset of training software. *PC Week*, 20–22.

Brown, J. (Dec. 1986). Chemical safety: A hazardous materials handling program. In "Software Selector." *Pollution Engineering,* 14.

Clerc, J. M. (Nov. 1987). Use of computers in medical technology education. *Lab. Med.* **18** (11), 773–775.

Environmental Protection Agency. Hazardous chemical reporting: Emergency planning and community right-to-know programs. *Federal Register* **52**, 2836 (January 27, 1987).

Genium Publishing Corporation, 1145 Catalyn Street, Schenectady, NY 12303, (518) 377-8854.

Harris, R. E. (1983). An Experimental Study to Determine the Effectiveness of Computer-Assisted Instruction in Industrial Arts Wood Laboratory Safety. Doctoral dissertation. East Texas State University, Texarkana, TX.

Hazox, P.O. Box 637, Chadds Ford, PA 19317, (215 (338-2030.

Interactive Medical Communications, 100 Fifth Avenue, Waltham, MA 02154, (617) 890-7707.

J. T. Baker Company, 22 Red School Lane, Phillipsburg, NJ 08865, (201) 859-2151.

Joiner, R. L. (1982). Occupational health and environmental information systems: Basic considerations. *J Occupational Med.* **24**(10): 863–866.

Kushner, J. (May 1988). Computer-assisted instruction in hematology. *Lab. Med.* **14**(5): 307–309.

Lancianese, F. W. (1983). Computer-based safety training in action. *Occupational Hazards.* **45**(9), 54–57.

Levy, S. R. (July 1984). Industrial health education needs: A feasibility study. *JOM* **26**(7), 534–536.

Mentor Learning Systems, Inc., 1825 De La Cruz Boulevard, Santa Clara, CA 95050, (408) 988-4114.

Occupational Health Services, Inc., Suite 24007, 450 Seventh Avenue, New York, NY 10123, (212) 967-1100.

Occupational Safety and Health Administration, U.S. Department of Labor. Hazard Communication; Final Rule. *Federal Register* **48**(228), 53280 (November 25, 1980).

Occupational Safety and Health Administration, U.S. Department of Labor. Hazard Communication; Final Rule. *Federal Register* **52**(163)m 31852 (August 24, 1987).

Pennacchia, M. (Jan. 1987). Interactive training sets the pace. *Safety and Health,* 29–30.

Resource Consultants, Inc., 7121 Crossroads Boulevard, Nashville, TN 37024, (615) 373-5040.

Schaaf, D. (Ed.) (Oct. 1987). High tech vs. high touch. In "Computers-Based Training." *Training* (special report).

Thomas & Associates, 208 North Main Street, Woodsboro, MD 21798, (301) 898-5115.

Woolsoncroft, J. (Nov./Dec. 1987). Computer training maneuvers into the government. Government Data Systems, 22–23.

Sources of Computer Software

Two comprehensive published lists of MSDSs and chemical emergency planning software are available from:

Marsick, D. (May 1989). "Computer Resources and the Material Safety Data sheet," Occupational Safety and Health Administration, Washington, D.C.

USEPA (Sep. 1989). "Computer Systems for Chemical Emergency Planning", Office of Solid Waste and Emergency Response, Washington, DC (Pub. No. OSWER-89-005).

PART 2

CASE HISTORIES

CHAPTER 8

CHEMICAL STORAGE FOR INDUSTRIAL LABORATORIES

FRANK L. CHLAD
Laboratory Design Consultant
F.L. Chlad & Associates
Aurora, Ohio

8.1 INTRODUCTION: A CONSULTANT'S VIEW

The problem of adequate and safe chemical storage in research laboratories is a major concern. As a laboratory design and safety consultant I have had the opportunity to visit many industrial research facilities and have found a large majority of them to be extremely cluttered, underhooded, and arranged with vast quantities of chemicals stored in their work areas. To me these are conditions that have a very high potential for disaster.

Perhaps one of the best ways to illustrate what is an all too common situation in the storage of chemicals in an industrial setting is through examination of an actual case study.

A multinational firm employing over 600 scientists at one of its research facilities asked that I become involved as a consultant to survey its present chemical storage facilities and associated operating procedures and make recommendations as to their improvement.

The problems that were found were complex, and various factors had contributed greatly in bringing about the existing situation. A recent in-house survey had been taken at this facility which revealed widespread dissatisfaction with the chemical storage operation. Eighty-three percent of all the respondents to the survey reported that they had problems with the system then in use and listed the major reasons as follows:

The reorder system on staple items was inadequate.

Safe and proper storage methods were not followed or enforced.

There was a need for storeroom personnel to properly evaluate the condition of the chemicals on the shelves regularly and make proper disposition.

There was a lack of knowledge as to what was on hand, both in the main storage facility as well as in the individual research laboratories.

Scientists felt the necessity to maintain their own stocks of chemicals in their research areas, resulting in costly duplication, clutter, and very unsafe storage practices.

There was a total lack of any inventory system and no trained support staff to assist researchers.

From a space standpoint, there were two primary storage areas for chemicals at this facility. These consisted of a drum solvent and gas cylinder storage shed located outdoors and apart from the main complex, and a chemical storage room.

The chemical storage facility consisted of a single room approximately 40×20 feet in size. During my site visits the room was not well ventilated and strong chemical odors were present. Storage was effected by the use of metal shelving units that showed considerable age and corrosion. The units were much too high to be safe (approximately 10 feet) and had boxes stored on the top shelves, reaching a total height of about 13 feet.

Each of the bins contained an excessive amount of chemicals. These were not arranged by hazard class but rather grouped by alphabet heading. Thus, all the chemicals beginning with the letter A were put together on several shelves, which required users to physically sort through a large quantity of bottles to find what they were looking for. Needless to say, this arrangement was an incompatibility nightmare! In addition, none of the shelves had a front safety lip, and a good many of the bottles were actually hanging over the edge of the shelf, posing a very dangerous situation.

This storage area had an exceptionally high ceiling with exposed utility lines running through the room and a ventilation duct suspended below the utility lines. Whereas the room did have a fire detection device, it may well have been rendered useless because it was located on the ceiling with the ventilation duct directly below.

The room was equipped with a carbon dioxide fire suppression system— with no built-in delay to safely evacuate any personnel—and was fitted with explosion-proof lights and static-free light switches.

The most serious problem was that of the incompatibility of the chemicals stored within the room. Due to the woeful lack of adequate space allocated for storage, chemicals of all types and hazard class were stored together, an extremely unsafe condition.

The drum and cylinder storage building was approximately 50 x 60 feet. This building contained a large quantity of 55-gallon drums of various solvents that were stacked four and five drums high on metal racks, a large quantity of compressed gas cylinders, and a variety of chemicals staged here prior to disposal. From a physical standpoint the building appeared to meet minimum safety standards. However, since the building only had a roof and tarpaulin sides attached to a three-foot high concrete wainscoat, there were serious problems during hot summer months and very cold winter months. This is an "open access" area, and any persons needing any solvents or cylinders of compressed gas could simply come in and take what they needed, whenever they wanted. This, of course, meant that this area was only as safe as the *least* safety-conscious persons who utilized it. This operation could be greatly improved by designating someone to be responsible for this area and allowing only that trained individual to do the dispensing. Among the safety problems in this area were the lack of a dike arrangement to contain any spills, the use of spigots as opposed to the much safer rotary transfer pumps, the lack of drip pans with flame arresters, plus the dangerous practice of storing large quantities of gas cylinders in this area without regard to their compatibility.

The company that I have used as an example has corrected all of the problems and appears to be well on the way to having a safe and efficient operation. Recommendations that were acted upon include:

A complete and thorough inventory and evaluation of the entire chemical stock, ensuring that the ones that were outdated or contaminated were disposed of properly.

That all research groups agree upon a small number of basic chemicals that should be stocked. These chemicals should have a high turnover rate and be of the type that are common to all researchers regardless of their area of specialty.

That a proper computer inventory system be established so that all chemicals can be readily found, stock levels replenished, and users identified. During the normal workday someone must be physically present in the chemical stockroom to handle all sign-outs and provide assistance. During off hours a computer card or tag system should be put into effect so that materials taken by researchers when the stockroom is not attended are properly deducted from stock and the users identified. Researchers should have the right of free access to the stockroom when it is not attended; however, it must be stressed that with that right comes the responsibility of properly signing out materials so that the integrity of the inventory control system is not jeopardized. If the sign-out procedure does not live up to expectations, then the privilege of allowing off-hour access to the stockroom should be curtailed.

That a chemically oriented stockroom manager be hired as soon as possible.

To seriously consider having an in-house sales/service representative from a major chemical supply house. I have observed this operation in several major firms, and it appears to work quite well. One of the key features is that the supply house does the warehousing for other than the basic list of chemicals, thereby alleviating the need for large amounts of storage space. Firms utilizing this particular system report that they are well pleased and that it has eliminated much of the previous purchasing problems, such as exact pricing, returns, sourcing, back-orders, and tracing. Delivery has been excellent, with two- or three-day delivery being the norm.

In addition to adopting all the above recommendations except the last, the company has built a modern chemical storage facility complete with hazard class segregation and all the necessary safety features. The gas cylinders are now stored in a special separate facility, apart from the solvent drum stock, and the solvent storage area is properly diked and utilizes rotary pumps and drip pans with flame arresters. Perhaps one of the more significant improvements is the hiring of a well-qualified chemical storeroom manager who is both safety-conscious and service-oriented. One of his first accomplishments was to establish a computerized inventory control system.

8.2 STORAGE AND OPERATIONS PROCEDURES

Ideal chemical storage should be high on the list of priorities for any industrial research facility. This entails complete isolation of each hazard class and, in some cases, isolation of materials within the same class. In actual practice, however, such ideal storage is rare. From a practical standpoint, it is often necessary to group items so that whatever space is available is used in the safest manner. An important factor that must be stressed is that there are no absolutes in designing safe storage facilities. What is adequate for one operation may not be proper for another. The design has to be customized to suit the individual needs of each particular concern. We all operate under various constraints—the amount of space we have available, the size of our operation, the total dollars allocated to us. The best we can hope to accomplish is to carefully prioritize our safety needs and spend the safety dollars as wisely as possible.

The National Fire Protection Association Booklet No. 45, *Fire Protection for Laboratories Using Chemicals,* is an excellent resource for determining maximum amounts of flammable and combustible liquids in laboratory units.

Tables 8.1–8.4 are from NFPA No. 45 and illustrate various standards for quantities of chemicals in a laboratory, the classification of a laboratory based on construction, and fire protection and maximum allowable size of containers.

Fifty-five-gallon drums are commonly used to ship flammable liquids but are not intended as long-time inside storage containers. It is not safe to dispense from sealed drums exactly as they are received. The bung should be removed and replaced by an approved pressure and vacuum relief vent to protect against internal pressure buildup in the event of fire or if the drum is exposed to excessive heat or direct sunlight.

Ideally, drums should be stored on metal racks with the end bung openings toward an aisle and the side bung openings on the top. The drums, as well as the racks, should be grounded with a minimum length of American gauge 10

TABLE 8.1 Maximum Quantities of Flammable and Combustible Liquids in Laboratory Units[a] Outside of Approved Flammable Liquid Storage Rooms

Laboratory Unit Class	Flammable or Combustible Liquid Class	Excluding Quantities in Storage Cabinets and Safety Cans (gallons)			Including Quantities in Storage Cabinets and Safety Cans (gallons)		
		Maximum Quantity[b] per 100 Square Feet of Laboratory Unit	Maximum Allowable Quantity[c] per Laboratory Unit		Maximum Quantity[b] per 100 Square Feet of Laboratory Unit	Maximum Allowable Quantity[c] per Laboratory Unit	
			Unsprinkled	Sprinklered[e]		Unsprinkled	Sprinklered[e]
A	I	10	300	600	20	600	1200
	I, II, and IIIA[d]	20	400	800	40	800	1600
B	I	5	150	300	10	300	600
	I, II, and IIIA[d]	10	200	400	20	400	800
C	I	2	75	150	4	150	300
	I, II, and IIIA[d]	4	100	200	8	200	400

[a] Class A laboratory units shall not be used as instructional laboratory units and the maximum quantities of flammable and combustible liquids in Class B and Class C instructional laboratory units shall be 50% of those in Table 8.1.

[b] For maximum container sizes see Table 8.3.

[c] Regardless of the maximum allowable quantity, the maximum amount in a laboratory unit shall never exceed that calculated by using the maximum quantity per 100 square feet.

[d] The maximum quantities of Class I liquids alone.

[e] In laboratory units where water creates a serious fire or personnel hazard, a nonwater extinguishing system may be substituted for sprinklers.

TABLE 8.2 Construction and Fire Protection Requirements for Laboratory Units[a]

Laboratory Unit Class	Area of Laboratory Unit (sq. ft.)	Nonsprinklered Laboratory Units				Sprinklered Laboratory Units[b]	
		In Fire-Resistive, Protected Noncombustible or Noncombustible Buildings		In Heavy Timber, Ordinary or Wood Frame Buildings		Any Building or Laboratory Construction	
		Separation from Non-laboratory Areas[c]	Separation from Laboratory Units of Equal or Lower Hazard Class[c]	Separation from Non-laboratory Areas[c]	Separation from Laboratory Units of Equal or Lower Hazard Class[c]	Separation from Non-laboratory Areas[c]	Separation from Laboratory Units of Equal or Lower Hazard Class[c]
A	Under 1000	1 hour	1 hour	2 hours	1 hour	1 hour	Noncombustible[d]
	1001–2000	1 hour	1 hour	Not permitted	Not permitted	1 hour	Noncombustible[d]
	2001–5000	2 hours	1 hour	Not permitted	Not permitted	1 hour	Noncombustible[d]
	5001–10000	Not permitted	Not permitted	Not permitted	Not permitted	1 hour	1 hour
	10001 or more	Not permitted	Not permitted	Not permitted	Not permitted	Not permitted	Not permitted
B	Under 20000	1 hour	Noncombustible[d]	1 hour	1 hour	Non-comb.[d,e]	Noncombustible[d]
	20001 or more	Not permitted	Not permitted	Not permitted	Not permitted	Not permitted	Not permitted
C	Under 10000	1 hour	Noncombustible[d]	1 hour	Noncombustible[d]	Non-comb.[d,e]	No requirement
	10000 or more	1 hour	Noncombustible[d]	1 hour	1 hour	Non-comb.[d,e]	Noncombustible[d]

[a]Where a laboratory unit contains an explosion hazard, appropriate explosion protection shall be provided for adjoining laboratory units and non-laboratory areas.

[b]In laboratory units where water creates a serious fire or personnel hazard, a nonwater extinguishing systme may be substituted for sprinklers.

[c]For a discussion of fire resistance of building materials, including the resistance of wall, partition, floor and ceiling construction, see *Fire Protection Handbook*, 13th ed., NFPA, 1969, pp. 8-86–8-124. For information on the fire resistance, installation, and maintenance requirements of Fire Doors, see NFPA 80-*Standard for Fire Doors and Windows*.

[d]May be ½-hour fire resistance rated combustible construction in lieu of noncombustible construction separation.

[e]In educational occupancies laboratory units shall be separated from nonlaboratory areas by construction having not less than one-hour fire-resistance.

TABLE 8.3 Maximum Allowable Size of Containers

Container Type	Flammable Liquids			Combustible Liquids	
	Class IA[a]	Class IB[b]	Class IC[c]	Class II[d]	Class IIIA[e]
Glass	1 pt.[f]	1 qt.[f]	1 gal.	1 gal.	5 gal.
Metal (other than DOT drums) or approved plastic	1 gal.	5 gal.[g]	5 gal.[g]	5 gal.[g]	5 gal.[g]
Safety cans	2 gal.	5 gal.[g]	5 gal.[g]	5 gal.[g]	5 gal.[g]
Metal drums (DOT spec.)	5 gal.[g]	5 gal.[g]	5 gal.[g]	60 gal.[g]	60 gal.[g]

[a] Class IA liquids are those having flash points below 73 °F (22.8 °C) and boiling points below 100 °F (37.8 °C).
[b] Class IB liquids are those having flash points below 73 °F (22.8 °C) and boiling points at or above 100 °F. (37.8 °C).
[c] Class IC liquids are those having flash points at or above 73 °F (22.8 °C) and below 100 °F (37.8 °C).
[d] Class II liquids are those having flash points at or above 100 °F (37.8 °C) and below 140 °F (60 °C).
[e] Class IIIA liquids are those having flash points at or above 140 °F (60 °C) and below 200 °F (93.4 °C).
[f] Sizes as large as 1 gallon may be used if needed and if the required liquid purity would be adversely affected by storage in metal or if the liquid would cause excessive corrosion of the metal container.
[g] In instructional laboratories, no container for Class I or Class II liquids shall exceed a capacity of 1 gallon, except that safety cans may be of 2-gallon capacity.
[h] See exception to 4 in Table 8.4.

TABLE 8.4 Storage

1. Hazardous chemical inventories in a laboratory's storage facility shall be within the prescribed capacities of the facility for the various kinds of chemicals.

2. The quantities of hazardous chemicals within each laboratory unit shall not exceed the permitted quantities in Chapter 2 and Table 8.1.

3. Hazardous chemicals in the open in laboratories shall be kept to a minimum necessary for the work being done.

4. Container types and maximum capacities shall comply with Table 8.3, except that Class 1A and Class 1B flammable liquids may be stored in glass containers of not more than one-gallon capacity if the required liquid purity would be affected by storage in metal containers or if the liquid would cause excessive corrosion of the metal container.

Exception: Drums of up to and including 60 gallons capacity are permitted in approved storage rooms.

5. Incompatible materials shall be segregated to prevent accidental contact with one another.

wire. Because effective grounding requires good metal-to-metal contact, be sure that all paint, dirt, and corrosion is first removed from the contact areas. Remember that it is also necessary to provide bonding to metal receiving containers to prevent accumulation of static electricity (which will discharge to the ground, creating a spark that could ignite the flammable vapors). Drip pans that have flame arresters should be placed under the faucets (Figure 8.1).

Dispensing from drums is usually done by open of two methods. The first is gravity-based through drum faucets that are self-closing and require constant hand pressure for operation. Faucets of plastic construction are not generally acceptable due to chemical action on the plastic materials. The second, and much safer method, is to use an approved hand-operated rotary transfer pump. Such pumps have metering options and permit immediate cutoff control to prevent overflow and spillage, can be reversed so that excess liquid can be siphoned off in case of overfilling, and can be equipped with drip returns so that any excess liquid can be returned to the drum (Figure 8.2).

No discussion of storage would be complete without mentioning the use of laboratory fume hoods to store large quantities of chemicals. I continue to be amazed (and alarmed) at the large number of research laboratories that have fume hoods so crammed-full of chemicals that reactions that should be conducted in the hood are run on open benches. In many instances, there are so many chemicals in the hood that air flow is blocked, as well as posing the additional hazard of the incompatibility of the chemicals.

I cannot stress too strongly that hoods should be utilized to run reactions and *not* to store chemicals. One excellent solution to the problem of storing chemicals that are malodorous or extremely toxic is to design the laboratory with an extra hood to be used only for that purpose. For example, in a laboratory designed for four persons, each person would have a hood in which to run reactions, and there would be one additional hood in which to store chemicals for all four persons. This is an energy-saving solution as well. If there were only four hoods, each researcher would store his chemicals in his own hood, necessitating all four hoods to be operating 24 hours a day. With the addition of one extra hood strictly for the storage of chemicals, the regular hoods are on/off as needed, and the only hood that must always be on is the one storage hood. In this manner the laboratory is much safer, and the cost of the additional hood is paid for in a few years through savings in the cost of energy (Figure 8.3).

The placement of laboratory fume hoods in the research area is a critical design consideration. Hoods should never be placed next to doorways or adjacent to high-traffic areas because of the wind currents that will be caused, which tend to draw fumes out of the hood, and also because of the obvious safety factors involving the blocking of exits and evacuation of personnel in the event of fire or explosion in the hood. A very good design rule to follow is *People by doorways, hoods by window walls.*

FIGURE 8.1 Proper dispensing for a 55-gallon drum. Note the grounded drum, bonding to receiving container, and use of drip pan with flame arrester.

FIGURE 8.2 The use of a rotary transfer pump allows metering and immediate cutoff.

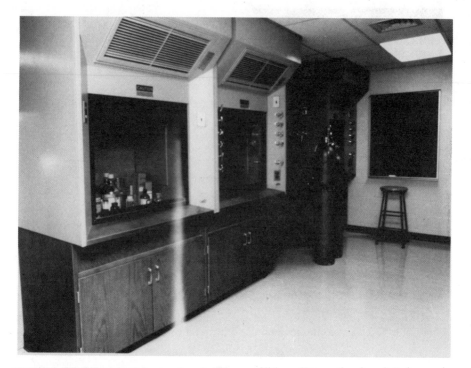

FIGURE 8.3   Chemical storage hood with no utilities; allows other hoods to be used for running reactions.

8.3   PEOPLE AND CHEMICAL STORAGE

Of paramount importance when considering the matter of industrial storage of chemicals is a thorough examination of the open-access versus controlled-access chemical stockroom.

Chemical storerooms should be conveniently located and open during what would be considered normal working hours (8:00 am–5:00 pm) so that laboratory workers need not store excessive quantities of chemicals in their laboratories and can readily obtain chemicals and glassware for their projects. This, however, does not imply that all laboratory workers should have un-limited access to chemicals 24 hours a day, seven days a week. Procedures must be established for the operation of any chemical stockroom that places responsibility for its safety and inventory control in the hands of one person.

The chaotic situation that occurred at the facility I described earlier was due in large part to the fact that an open stockroom existed. This open concept has great appeal to scientists because it affords them free access to the shelves and they can pick up whatever they want (or are able to find!) without any burden-some record-keeping or bureaucratic red tape. The very fact that a person can come and take something whenever they need it is the very reason that this open concept always fails. It fails because no controls exist to ensure proper replenishment of stock, an accurate knowledge of what is on hand, and a sense of order and safety on the shelves.

For these reasons, a lack of confidence in the entire stockroom operation builds up over a period of time, and research groups then feel that they must fend for themselves, never realizing that *they* were the major factor in the demise of the stockroom. The next step in this evolution is that each research laboratory feels the need to have its own ministockroom and become completely independent of any centralized operation. This, of course, then creates the problems discussed earlier: that of an excess amount of chemicals stored in the research laboratory, cluttered and all too often incompatible storage conditions (particularly in the hoods), no real inventory control facility-wide, and ultimately a very costly waste disposal problem.

The controlled-access stockroom should not be viewed as a system that inhibits or restricts the activity of researchers, but rather as an additional service that is made available to make their task easier, quicker, and more efficient. Research personnel can still peruse the shelves. However, the added bonus would be the establishment of a technically trained person who could assist the researchers to better find what they are seeking, and, perhaps even more importantly, to record the activity. This is absolutely essential relative to ensuring proper inventory control. Typical complaints voiced by researchers include:

"Items are not reordered until we complain we're out."

"Many of the chemicals in the stockroom are out of date and need to be culled. . . . We don't have any system."

"I waste so much time looking for chemicals we don't have that I just order them and keep them locked up in my own laboratory."

"No one seems to know (or care) what we have, or where it is."

A proper computer inventory system would alleviate these problems and establish an excellent information source. Minimum stock levels could be established, past usage could be determined, and a record of where chemicals are located in the entire facility could be easily provided (a necessity now with SARA and right-to-know laws).

It is essential that the stockroom and receiving personnel be knowledgeable about and trained in the handling of hazardous substances. Several federal regulations now make it mandatory that all employees that handle chemicals receive proper training, training that is on-going and fully documented. Information must be provided about environmental and hazard-initiating exposures that must be avoided. Some of the more common items with which receiving room and stockroom personnel need to be familiar are:

Proper use of material handling equipment

Basic data entry and retrieval on a personal computer

EPA and DOT regulations pertaining to handling and storage of chemicals (including proper filing of MSDSs)

Knowledge of various safety codes, such as DOT labels on shipping packages and cartons, numbering system on compressed gas cylinders, and the NFPA diamond code

Emergency procedures to be followed in the event of a spill or fire

Knowledge of the incompatibility of some chemicals and the sensitivity of some items to water, light, heat, and other hazards

All too often it is assumed that a storeroom person is nothing more than an entry-level worker who requires a strong set of shoulders to lift cartons and move cylinders. This kind of thinking can be very short-sighted and bring about near-disasters in the storeroom or receiving areas. What is needed is an individual who is safety conscious and service oriented, has some familiarity with chemistry, can keep accurate records on a computer, and tries to be genuinely helpful to those seeking assistance. Even though you may have to pay a slightly higher salary, seeking a more technically trained individual for this type of position will more than pay for itself in what can be provided as support service for researchers in the organization.

8.4 SAFETY EQUIPMENT FOR CHEMICAL STORAGE

Fire-protected facilities are essential for the safe storage of flammable liquids. Chapter 3 gives recommendations for storeroom construction and capacity for storing large amounts of flammable liquids; however, there will still be other areas where limited quantities are kept. Among those areas are safety cans, flammable liquid storage cabinets, and specially designed refrigerators for the purposes of limited quantity storage.

Safety cans are containers that have built-in safety features for protecting flammable liquids from exposure to a fire situation. In a fire situation, a safety can is exposed to extremely high temperatures. The heat is transmitted to the contents, which, in turn, boil and produce a large vapor pressure. Every safety can is fitted with a spring-loaded cap that vents the vapors without bursting the safety can. The other safety feature of a safety can is the flame arrester, which is a cylindrical wire screen. Vapors emitted from a safety can will ignite when exposed to the flames of a fire. Since flames usually flash back to the source of liquid, the flame arrester serves as a heat dissipator. The temperature in the space above the liquids in a safety can is lowered below the ignition temperature, and ignition of the contents is eliminated. Figure 8.4 shows a cutaway view of a safety can that reveals the positioning of a flame arrester.

Safety cans are available in several sizes, the largest of which has a 5-gallon capacity. Table H-12 in OSHA regulation No. 29 CFR 1910:106 shows the maximum allowable size of safety cans for the various classes of flammable and combustible liquids. Occupational Safety and Health Administration Code 1910:106(d)(2) requires that only *approved* containers (except those in DOT-Spec metal drums) are to be used to store flammable liquids and combustibles. Many safety cans commercially available have been tested and approved by Underwriters Laboratories (UL) and/or Factory Mutual System (FM). In addition, OSHA regulations 29 CFR 1910.144(a)(ii) require safety cans to be painted red and carry either a yellow band or have the name of the contents painted or stenciled in yellow on the can for flammable liquids with a flash point at or below 80°F.

Flammable liquid storage cabinets are designed to keep the temperature at the top center of the cabinet interior below 325°F when subjected to a 10-minute fire test. (*Note:* A storage cabinet for flammable liquids is not fireproof but only protects the contents from extreme temperatures for a limited time to allow evacuation of personnel and prompt entry of the fire fighters.)

Cabinets built to withstand the temperature rating during the 10-minute

FIGURE 8.4   Cutaway view of safety can reveals position of the cylindrical flame arrester. (Used with permission of the Justrite Manufacturing Company, Des Plaines, IL.)

FIGURE 8.5 A metal storage cabinet approved for storing flammable liquids. Such cabinets are constructed with double metal walls, separated by a 1½-inch air space. (Used with permission of the Justrite Manufacturing Company, Des Plaines, IL.)

fire test (prescribed in NFPA 251-1969) are acceptable by OSHA standards if (a) the maximum capacity of Class I and II liquids is not more than 60 gallons (or more than 120 gallons for Class II liquids) and (b) the cabinet is labeled with conspicuous lettering: FLAMMABLE—KEEP FIRE AWAY.

Construction requirements for wooden and metal cabinets acceptable for the storage of flammable and combustible liquids are given in 29 CFR 1910.106(d)(3). Figure 8.5 shows a picture of a double-walled metal storage cabinet for flammable liquid storage.

For small quantities of Class IA liquids (which have low boiling points), refrigerated storage may be necessary to prevent volatilization. Special refrigerators that may safely store flammable liquids have a spark-free interior in that all wiring and thermostat controls have been removed from the in-

FIGURE 8.6   An explosionproof refrigerator. (Used with permission of Marvel, Inc., Dayton, OH.)

terior. Two types of these refrigerators are commercially available: a flammable liquid storage model and an explosion-proof model. A flammable liquid storage model is normally used in a nonexplosive area where no flammable vapors are present. Such a refrigerator is normally powered through a standard three-wire cord plugged into an electrical outlet. An explosion-proof refrigerator (Figure 8.6) is required when the area in which the refrigerator will be located has the potential for ignition of flammable vapors. An explosion-proof refrigerator is supplied with a "pigtail" cord that must be wired directly to a power source using metal conduit as specified by local electrical codes. Choosing the appropriate refrigerator will depend on the area in which it will be located.

In addition to the above, the safe storage of chemicals also requires that a

FIGURE 8.7 An automatic carbon dioxide fire suppression system.

FIGURE 8.8 Plastic explosion blow-out skylight in main chemical storage facility.

facility be properly designed and equipped with safety features such as an adequate fire suppression system (Figure 8.7), static-free switches and outlets, and explosion-relief capability such as plastic blow-out skylights (Figure 8.8).

8.5 CONCLUSION

We must start making a serious and dedicated commitment to safety. We must ensure that safety is designed into our chemical storage facilities and into our daily operations. We must continually look for opportunities to promote safety and to convince top management that safety, as expensive as it sometimes is, is worth every penny spent on it. We must continually evaluate our procedures and systems and ensure that we are operating in the safest manner possible.

The great scientist Louis Pasteur expressed it exceedingly well when he said, "Take interest, I implore you . . . in those sacred dwellings one designates by the expressive term . . . Laboratories. . . . Demand that they be multiplied, and that they be well-adorned. These are the temples of the future . . . temples of well-being and happiness . . . and it is there that humanity grows greater, stronger and better."

We must expend the time, energy, and dollars that are required to provide a safe environment in which to work. We really cannot afford to do any less.

CHAPTER 9

IMPLEMENTATION OF AN ON-LINE IMS DATABASE SYSTEM FOR WAREHOUSE AND INVENTORY MANAGEMENT

E. LAMAR HOUSTON
Director, Research Services
University of Georgia
Athens, Georgia

9.1 INTRODUCTION

The University of Georgia organized a facility called Central Research Stores (CRS) during 1967. The purpose of this facility was to save the university money through volume purchases as well as to have a ready source of supply of scientific apparatus and chemicals. Construction was finished on a 20000-square-foot building during the fall of 1969, and shortly thereafter the operation was in full swing. All administrative functions were executed manually for nine years, but after that time sales passed the two million dollar mark. The operation was beginning to drown in paper work. The need for a computer was evident. Direct access, on-line capability was desired, so CRS staff began talking and visiting commercial firms to locate a software package. It was discovered that there were no inventory software packages available, either because of price or because companies did not want to give out proprietary information. Therefore, university analysts and programmers began developing in-house software.

Manual operation ceased and computer operation began on March 19, 1979. After considerable transition difficulty, the computer operation smoothed out during the summer of 1979, and today it is functioning in an impressive manner. Some development work still remains, and periodically changes are necessary to some of the programs to give more flexibility.

9.2 GENERAL SYSTEM GOALS

At the outset of planning for the new database system, there were several major operating components of the CRS facility that would be automated by this new system:

1. An inventory management system that would provide timely information about the inventory contained in the warehouse of the CRS facility

2. An implementation of the basic purchasing function, including constructing and issuing purchase orders and following through on the status of purchase orders once they have been placed with vendors

3. A more extended sales function, including the order entry process, commitment of stock from the warehouse, and tracking goods as they progress through the various steps of the sales process

4. An interface to the then-existing manual accounting system of CRS, with an eventual progression to an automated accounting database application

5. An easy interface to the University of Georgia accounting procedures by means of automated analysis of both the purchasing and sales functions in the CRS system

It was also decided that the tools for implementing this particular application would be IBM's Information Management System (IMS). Also, the system was to be an on-line and real-time application so that stock information displayed on terminals would reflect actual real-time stock balances within the warehouse. The large mainframe at the university's computer center and the IMS database system would be used with remote inquiry, update, and printing at the CRS location. This would take advantage of the staff already available and eliminate the addition of data processing staff at CRS that would have been needed by selecting an in-house minicomputer. And finally, it was decided that the system should be recoverable and secure, both of which are attributes adequately provided by the IMS software package.

Figure 9.1 represents a map of the structure of the CRS databases. The databases are connected in a logical manner that allows browsing from one database to another. All one must do is key in the segment to reach a specific database. For example, to move from the vendor database (VN) to the customer database (CU) one must key in the segment CU. One can also reach different levels of each database by keying in the various segments listed on the map.

9.3 INVENTORY PROCESSING AND INFORMATION

The original goals for this particular phase of the application were as follows:

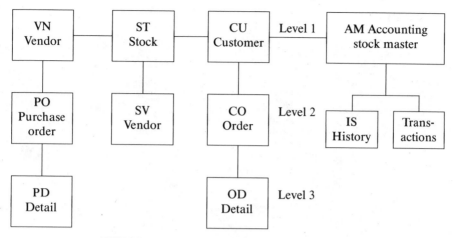

FIGURE 9.1   Central Research Stores databases.

1. To maintain a stock master with information about specific items, including the description, issue units, purchase units, quantity on order, quantity on hand, quantity backordered, and other statistics about each item
2. To automatically maintain receipts and issue information about each item
3. To provide varied lists for taking inventory and analyzing stock items
4. To generate an automated sales catalog with camera-ready proof
5. To generate any miscellaneous stock-related reports as required on an ad hoc basis
6. To classify stock items by type, markup, and other attributes
7. To provide on-line inquiry of all customer orders and purchase orders for a given stock item

Table 9.1 is an example of a Stock Database Inquiry Update Screen. Contained on this screen are the basic data elements that are relevant to decision-making about a given stock item. As can be seen, various categories of data elements are maintained about each stock item, for example, material types, the unit of ordering for purchasing purposes, the case quantity or multiplier of number of issue units per order unit, the average unit cost, and previous average unit cost. Also, information about stock availability is readily obtainable by looking at this screen. History information about period issues and receipts, demand, stockouts, and backorders are also readily accessible.

The previously discussed screen, as well as many others, was implemented using IBM's IMS software package as well as their Application Development Facility (ADF) package. ADF has proven to be a very effective productivity enhancer at the University of Georgia, allowing on-line IMS applications to

TABLE 9.1

```
                    STOCK DATABASE INQUIRY/UPDATE                      08/16/89
                                                                      15:28:45
STOCK NUMBER   102900      ST COMMODITY   2899735
DESC           ACETONE AR CH32CO
               FLAMM. LIQ. 301-X          UN 1090
MATERIAL TYPE  11             CHEMICAL EXP           COST CODE    2
STOCK TYPE     1              REGULAR                UNIT COST        6.0407
ISSUE UNIT     4L             CASE QTY         1     PREV COST        5.7540
ACTIVE CODE                   ORDER UNIT     4L
ON HAND              17       SUSPENSE         0     15      %        6.9518
COMMITTED             0       AVAILABLE       17     BOOKSTORE        7.7263
ON ORDER              0       BACK ORDER       0
LAST ISSUE     08/11/89       LAST RECEIPT 08/01/89
PERIOD ISSUES      1380       YEAR TO DATE     1380  FIXED QTY        1
RECEIPTS           1346                        1346  STOCK MIN       12
DEMAND             1380                        1380  STOCK MAX       24
STOCK OUTS           11                          11
BACKORDERS           53                          53
OPTION:        TRX: 5ST    STOCK NUMBER: 102900
     *** ENTER DATA FOR UPDATE ***
```

be implemented in one-fourth to one-twentieth the time of a pure COBOL or PL/I IMS application. By using ADF for on-line inquiry an update in conjunction with Informatic's MARKIV for batch processing, significant productivity enhancements have been obtained over the past several years, allowing rapid implementation of systems similar to this particular application.

Other screens have been developed, which are not displayed in this chapter, that allow a CRS staff member to inquire into all the outstanding customer orders by customer number, as well as purchase orders by vendor number outstanding for a given stock item. Along with this information are provided various status codes to tell the status of each customer or purchase order in its normal life span. An additional screen was also provided to show all vendors who provide a given stock item, along with the representative for that vendor, the vendor catalog number, and the telephone and address of the vendor. This information can be used in constructing purchase orders.

Currently, the normal "reorder at minimum, reorder up to maximum" procedures are being used for replenishing stock at CRS; however, when the original CRS system was implemented, the data elements were provided in the stock segment of the database to be able to easily interface with IBM's INFOREM package. This package is a very sophisticated inventory forecasting and management package that has been purchased by CRS for later installation with the IMS inventory management system. The data elements for each stock item are already stored in the database. The basic functions of INFOREM, when completely installed, will be to:

1. Monitor and measure the average current demand for items at the stock-keeping unit level.
2. Produce forecasts of future stockkeeping unit demand.
3. Develop ordering decision rules for the optimum restocking of items.

4. Maintain accurate leadtime for vendors at the time level.

5. Simulate performance for a month, season, or year.

9.4 PURCHASING

The original goals for the purchasing component of this system were to:

1. Provide real-time inquiry and update information and purchase orders for each vendor.
2. Provide vendor cross-reference lists.
3. Provide real-time information about all orders for a stock item.
4. Control the status of a purchase order at the detail line level; in other words, be able to accommodate the partial completion, partial payment, and partial receipt of specific lines or parts of lines on a given purchase order.

Several screens were developed to fulfill the on-line inquiry requirement of this chapter. A Vendor Database Inquiry Update Screen was designed to show the vendor number, name, status code, payment terms required, shipping terms, outstanding contracts that might exist for a vendor, the vendor representative and phone number, and purchase from and pay to addresses. By using ADF's automatic segment browsing capability, a CRS staff member can easily browse through all outstanding purchase orders for a given vendor. By proceeding to the second level of the database, this staff member can also browse through all the detail lines for a given purchase order, showing quantities ordered and received, the unit price, vendor reference number, and the particular stock number associated with the detail line (Table 9.2). At the detail or line level of a purchase order (Table 9.3) data elements are carried to show

TABLE 9.2

```
          S E C O N D A R Y   K E Y   S E L E C T I O N
UPDATE            TRANSACTION: VENDOR
OPTION:      TRX: 5PD    KEY: 04700CRS-2026-A00000
SELECTION:            *** ENTER A SELECTION NUMBER FROM THIS SCREEN ***
                      QUANTITY   QUANTITY      UNIT    VENDOR      STOCK
          LINE  CHNG  ORDERED    RECEIVED      PRICE   REFERENCE   NUMBER
     1    001   01       1          1         174.44               9212006001
     2    002   01       3          3          83.35               9209040001
     3    003   01       1          1          63.00               9209039001
     4    004   01       2          2          30.25               9209037001
     5    005   01       1          1          98.56               9209036001
     6    006   01       1          1          23.75               9209036002
     7    007   01       1          1           9.38               9209014004
     8    008   01       1          1          46.25               9209022001
     9    009   01       1          1          66.53               9209022002
    10    010   01       3          0           3.00               9209022003
    11    011   01       1          0          77.53               9209022004
```

TABLE 9.3

```
                    VENDOR DATABASE - PURCHASE ORDER DETAIL           08/16/89
                                                                      10:10:20
PURCHASE ORDER      CRS-2026-A
LINE/DASH           002         01                  STATUS   1
STOCK ITEM          9209040001                      PUR/ISS INDIC   P
DESCRIPTION         P453,C18 EXTRACTION COLUM  SPECIAL ORDER ITEM

QUANTITY ORDERED         3                      QUANTITY RECEIVED          3
ORDER UNIT PRICE       83.35                    PURCH UNIT PRICE       83.35
VENDOR PACK SLIP    054660                      VENDOR INV REF NO
VENDOR PYMT REF#                                ORDER DATE          07/31/89
                    --------- STATUS ---------      --- DATE ---
RECEIVING                1         RECEIVED          08/04/89
INVCD BY VENDOR          1         INVOICED          08/10/89
PAID TO VENDOR          0         NOT PAID            /  /
OPTION:           TRX:   6PD
NEXT PO DETAIL NUMBER:   04700CRS-2026-A00201
```

the actual item ordered from the vendor, as well as the quantity ordered and received, the order unit price and purchase unit price, the vendor's packing slip number, vendor payment reference numbers, and the original order date. The various statuses of a given detail line, that is, the receiving, invoicing, and payment status codes, are carried at the detail line level reflecting whether or not each of these particular stages has been reached and the date on which the stage was reached for that line. Several screens have been provided to allow the CRS staff to inquire into a given purchase order and examine the status codes thereof.

Figure 9.2 shows the general flow of purchase order construction. The normal steps involved in this process are:

1. Determine the items to be ordered.
2. Obtain prices for items.
3. Select vendors for the items.
4. Enter the purchase orders into the system through the CRT.
5. Print purchase orders on the printer.
6. Confirm the purchase order as being issued.

Until the point when the purchase order is actually confirmed as issued, the CRS staff member can modify the purchase order contents in the database as necessary. After purchase order issue is confirmed, all changes to the purchase order are treated like change orders, and consequently account effects are generated in the accounting reports to reflect these as change orders or purchase order adjustments.

For the purchasing staff to know what items to reorder, a Stock Items Below Minimum Level Report is generated showing by vendor the necessary stock items required of that vendor, as well as quantities required to replenish the stock level in the warehouse. The actual entry of the purchase order involves calling up a blank screen and entering the vendor number, purchase order

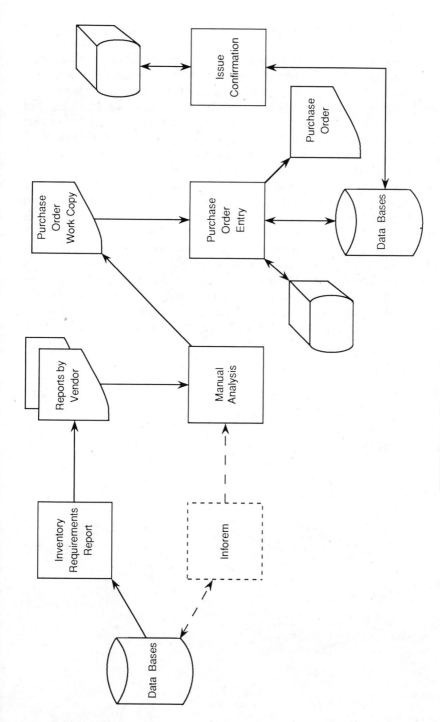

FIGURE 9.2 The purchase order function.

number, and for each line on the purchase order, the stock number, quantity ordered, and the price at which the quantity is being ordered. The person entering this purchase order has the option of taking the default purchase unit or specifying that the stock item is being ordered by issue units. The ability to specify that a given line on a purchase order is either in purchase or issue units allows the receiving of partially damaged cases at receipt time. Upon entry of the screen with the previously mentioned data elements, the purchase order is stored in the Vendor Purchase Order Database and a confirmation is returned to the screen. This confirmation also contains the purchase unit and the various status codes for the purchase order (Table 9.4).

By using the Purchase and Check Request Attachment Screen (Table 9.5), the CRS staff member can then request that the purchase order, which had just been entered, be printed on the remote 3287 printer. On the purchase order printing screen, the vendor number and purchase order are entered along with selection criteria: whether only received or not received items, invoiced or not invoiced items, and paid or not paid items should be printed. The option of whether to extend prices is also provided, as well as the ability to specify an alternate printer. When the purchase order is printed (Table 9.6), it is printed on 8½ × 11-inch white unlined paper and is in a suitable form to attach directly to the University Purchase and Check Request form. This is sent directly to the vendor.

TABLE 9.4

```
VENDOR 04700    PO NUM CRS-2026-A   FUNC    REF#                          89/08/16
OK
LN   CH   STOCK      QTY      UNIT      COST    FREIGHT I/P    RECV   S  R  I  P
001  01  9212006001    1   EA         174.44            P       1    1  1  0  0
002  01  9209040001    3   PKG         83.35            P       3    1  1  1  0
003  01  9209039001    1   EA          63.00            P       1    1  1  0  0
004  01  9209037001    2   EA          30.25            P       2    1  1  1  0
005  01  9209036001    1   EA          98.56            P       1    1  1  1  0
006  01  9209036002    1   EA          23.75            P       1    1  1  1  0
007  01  9209014004    1   EA           9.38            P       1    1  1  1  0
008  01  9209022001    1   CS          46.25            P       1    1  1  1  0
009  01  9209022002    1   CS          66.53            P       1    1  1  1  0
010  01  9209022003    3   EA           3.00            P       0    1  0  0  0
011  01  9209022004    1   CS          77.53            P       0    1  0  0  0
```

TABLE 9.5

```
                    C E N T R A L   R E S E A R C H   S T O R E S
                  PURCHASE AND CHECK REQUEST ATTACHMENT
DATE                  08/16/89
VENDOR NO             04700
PURORD NO             CRS-2026-A
VPSL NO
                                  ENTER X BELOW TO SELECT BOTH OPTIONS
      RECEIVED        0            0=NOT RECV       1=RECV
      INVOICED        0            0=NOT INVOICED   1=INVOICED
      PAID            0            0=NOT PAID       1=PAID
                      Y            EXTEND PRICES    Y OR N
                      CRSVP1       PRINTER - USUALLY CRSVP1
```

TABLE 9.6

THE UNIVERSITY OF GEORGIA	REQUEST NO CRS-2026-A
PURCHASE AND CHECK REQUEST ATTACHMENT	(04700)
	PAGE NO 1 OF 1 PAGES

ITEM NO	DESCRIPTION & SPECIFICATIONS	QUANTITY UNIT	INTERNAL USE	UNIT PRICE	TOTAL PRICE
001-01	9212006001 P1250 RIBOPROBE II CORE	1 X EA		174.44	174.44
002-01	9209040001 P453,C18 EXTRACTION COLUM	3 X PKG		83.35	250.05
003-01	9209039001 OB1312-52 ALKALINE PHOSPH	1 X EA		63.00	63.00
004-01	9209037001 14-649-5 DIGIT. STOPWATCH	2 X EA		30.25	60.50
005-01	9209036001 15-077-8 THERMOMETER	1 X EA		98.56	98.56
006-01	9209036002 14-511-59 STIR BAR KIT	1 X EA		23.75	23.75
007-01	9209014004 07-905 HAND HELD COUNTER	1 X EA		9.38	9.38
008-01	9209022001 13-070-3 INOCULATING LOOP	1 X CS		46.25	46.25
009-01	9209022002 13-675-24B DISP. PIPETS	1 X CS		66.53	66.53
010-01	9209022003 03-621B TUBE BRUSHES	3 X EA		3.00	9.00
011-01	9209022004 13-675-24C DISP. PIPETS	1 X CS		77.53	77.53
	PO TOTAL =				878.99

Receiving goods consists of the following major steps:

1. Recording the receipt of orders in the computer system
2. Recording confirmation of payments to the vendor by the University of Georgia's Accounts Payable Department
3. Providing reports at each of the preceding stages
4. Accommodating change orders, either before or after receipt of goods from the vendor
5. Allowing other adjustments to purchase orders
6. Handling freight charges, which are added to a specific purchase order, and the redistribution of these changes, if necessary, back to the original ordering customer

Figure 9.3 shows how receiving reports and changes in amendments to orders are entered through the terminal into the Vendor/Purchase Order Database. Various vendor analysis reports are provided from this database and automatic relief programs are run overnight to scan the databases and automatically print picking and packing slips the following morning to relieve backorders. In addition to various backorder relief reports, this feature eliminates the need for the CRS staff to constantly examine receiving reports to see which backorders can be relieved.

9.5 SALES PROCESSING

The customer order entry phase of this system includes the following major functions:

1. Recording the original customer orders
2. Printing picking and packing slips to go to the warehouse
3. Adjusting inventory and customer orders, as necessary, based on reconciliation with the actual stock in the warehouse
4. Shipping the goods to the customer by freezing the price in the database, and so on
5. The automatic generation of an invoice to be sent to the customer at billing time
6. Handling multiple generations of a customer order as previously nonavailable items are received through backorder processing

Figure 9.4 shows the normal flow of the customer sales function. Written orders or telephone calls are entered through the terminal to the customer database, picking/packing slips are sent to the warehouse, and then upon confirmation that the goods have been shipped or delivered, the database is updated through the terminal and an invoice is automatically printed for filing by customer number.

Several screens were provided for inquiry into the availability and prices of several stock items at one time (Table 9.7). This allows immediate access to the information in response to telephone inquiries or written orders. The Customer Order Entry Screen can also be used for inquiring into the status of a specific customer order (Table 9.8). A customer order is actually entered by calling up a blank Customer Order Entry Screen and entering the customer number, order number, a shipping address if different from the normal billing address, and the detail lines for the customer order. Each line requires only a stock number and a quantity ordered. The computer system automatically checks stock available and splits lines that are partially available into items to be shipped and items to be backordered. The average unit price, currently available in the stock database, is frozen for items to be shipped and extended prices are automatically calculated.

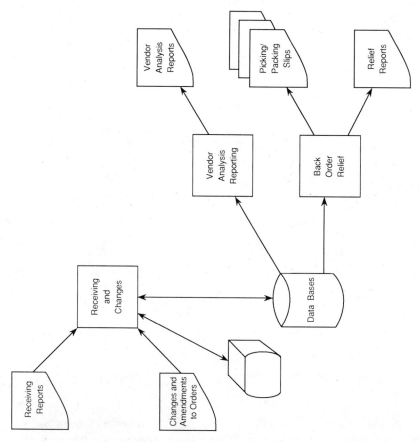

FIGURE 9.3 Receipt of goods.

TABLE 9.7

```
102900                 4L    |  106110                 4L
   ACETONE  AR  CH32CO        |     ACETONITRILE  HPLC
                             |
   UNIVERSITY PRICE  $   6.95 |     UNIVERSITY PRICE  $   25.70
   BOOKSTORE   PRICE  $   7.71 |     BOOKSTORE   PRICE  $   28.53
   COST              $   6.04 |     COST              $   22.35
   AVAILABLE QTY ====>    17  |     AVAILABLE QTY ====>    27
===============================|===============================
110500                 4 LB  |  191500              500ML
   ALCONOX                    |     BUFFER SOLUTION PH7
                             |     YELLOW
   UNIVERSITY PRICE  $   6.14 |     UNIVERSITY PRICE  $    1.71
   BOOKSTORE   PRICE  $   6.82 |     BOOKSTORE   PRICE  $    1.90
   COST              $   5.34 |     COST              $    1.49
   AVAILABLE QTY ====>    89  |     AVAILABLE QTY ====>    45
===============================|===============================
164820                 4L    |  234300                 4L
   BENZENE  AR  C6H6          |     CHLOROFORM  AR
                             |     CHCL3
   UNIVERSITY PRICE  $   6.44 |     UNIVERSITY PRICE  $   11.24
   BOOKSTORE   PRICE  $   7.15 |     BOOKSTORE   PRICE  $   12.48
   COST              $   5.60 |     COST              $    9.77
   AVAILABLE QTY ====>     7  |     AVAILABLE QTY ====>    39
```

TABLE 9.8

CUSTOMER 00121	ORDER 9156011	FUNC		SHIP TO GEORGE BAUGHMAN					
REF	DASH	CONF		TMI					
89/08/16				DAWSON				RM 367	
OK									

LINE	DASH	STOCK	ORDER	BO	SHIP	UNIT PRICE	S	I	P	DASH
001	01	9156011001	1		1	13.17	2	1	1	006
002	01	830305	1		1	8.24	2	1	1	001
003	01	9156011003	1		1	7.10	2	1	1	004
004	01	9156011004	1		1	6.73	2	1	1	002
005	01	9156011005	1		1	72.11	2	1	1	007
006	01	9156011006	1		1	4.75	2	1	1	005
007	01	726800	1		1	30.50	2	1	1	001
008	01	509900	1		1	12.40	2	1	1	001
009	01	9156011009	1		1	5.12	2	1	1	004
010	01	9156011010	1		1	3.16	2	1	1	003
011	01	848300	10		10	1.01	2	1	1	001
012	01	SFREIGHT	183		183	0.01	2	1	1	008
013	01	SFREIGHT	183		183	0.01	2	1	1	008

The sales clerk then can use a Picking/Packing Slip Invoice Printing Screen (Table 9.9) to enter the customer number and customer order number, and request any given generation picking/packing slip for a given customer order. Also available on this screen are the options previously mentioned for the purchase order printing screen. That is, the clerk has the option of just getting a normal picking/packing slip or selectively printing a customer order showing only lines that have been picked and packed but not shipped, or only the lines that have been shipped, or only the lines that have been backordered. Also, by selecting a different status code, items either invoiced or not invoiced can be printed. Other options are whether to extend prices and the specification of an alternate printer.

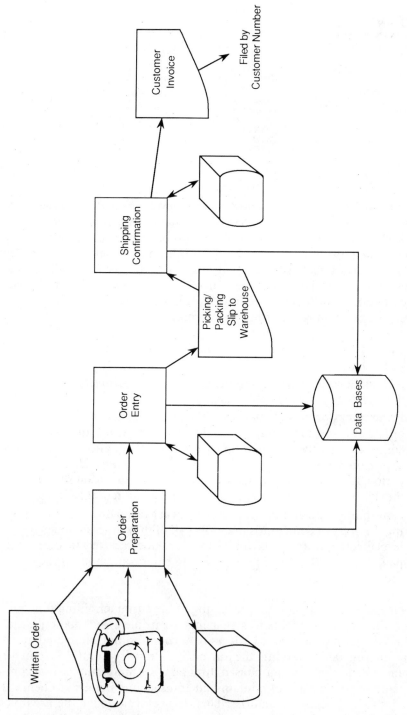

FIGURE 9.4 The customer sales function.

TABLE 9.9

```
                     C E N T R A L   R E S E A R C H   S T O R E S
                     PICKING/PACKING SLIP - INVOICE PRINTER
DATE                 08/16/89
CUSTOMER NO          00121
ORDER NO             9156011
PICKING SLIP#:       002
   PICKED/SHIPPED    1          0=NOT PP    1=PP NOT SHPD   2=SHIPPED    3=BACKORD
   INVOICED          X          0=NOT INV   1=REG INV       2=PRO FORMA
   PAID FOR          X          0=NOT PAID 1=PAID
EXTEND PRICES?       N          N Y I OR P
PRINTER TERMINAL     CRSVP2     USUALLY CRSVP2
RESPONSE
```

We plan to implement as our next stage of development direct order entry by our customers. This will be a paperless order entry system whereby customers who have access to the mainframe will be able to call up order entry screens in their own offices, enter the order information, and have the order print out in our offices. This process will ensure that the customers will receive items faster and, at the same time, our offices will save on labor charges because customers will enter their own orders for us.

The customer billing function has four major subfunctions:

1. Recording non-university payments and associated control information.
2. Automatic collection or reimbursal, if necessary, to university customers through the university accounting system.
3. Printing customer statements on demand with the ability to freeze orders at cutoff for billing processing. In other words, the system can continue to receive orders from customers without affecting what was shown on the statements as they were originally at the end of the billing period. Rerunning statements will produce the same results because of the date effective logic used in the implementation of this function.
4. Controlling the status of customer orders at a detail line level, yet allowing the entry of commands that affect all of these lines at an aggregate level. In other words, given lines can be ordered, received, invoiced, and paid, yet the CRS staff does not have to enter commands at the detail line.

Figure 9.5 shows the normal processing of the customer billing function. The databases are scanned and statements are printed. Automatic journal vouchers are generated to go to University Accounting and automated transactions are generated to go into the university's Property Control System for equipment items that were purchased through CRS. Simultaneously, a computer tape is generated that contains all transactions involving chemical deliveries to individual laboratories. This tape is forwarded to the Environmental Safety Services Department and becomes the basis of a hazardous

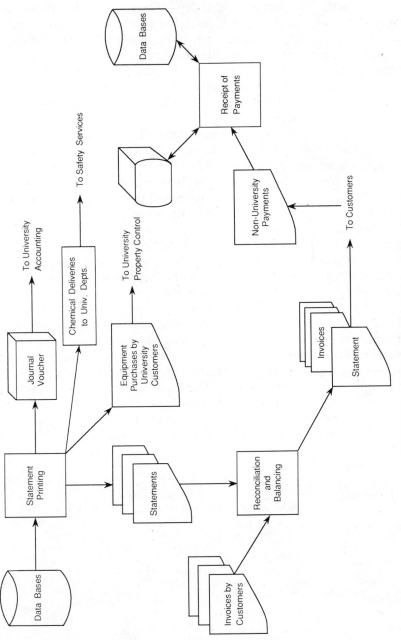

FIGURE 9.5 Flow of the customer billing function.

chemical inventory report sent to the Georgia Department of Labor. Georgia's employee right-to-know law specifies that periodic hazardous chemical inventory reports be forwarded to the state department of labor. The statements are taken and analyzed. Reconciliation occurs by comparing the originally printed invoices and making sure that the information in the invoices matches that on the statements. If necessary, statements can be reprinted. Upon the final completion of the statement printing, the invoices are bundled with the statements and shipped to the customers. No action is required of university customers at this point. The funds are collected automatically from their expense accounts. For non-university customers, however, payments are received and are entered through the terminal to indicate that the specific lines or entire orders for given customer orders have been paid.

At the detail customer order line level in the customer database (Table 9.10), data elements are carried to indicate what the specific order line represents as well as its status. Descriptive data elements include the stock number, quantity ordered, quantity shipped, the frozen unit price for this specific order line, the unit cost to CRS, and payment descriptive information, for example, the journal entry or ticket number for a university payment, if appropriate. The status codes, as in the purchasing function, are the shipping, invoicing, and payment status code showing whether the specific line has been shipped, invoiced, and/or paid, and the specific dates on which these stages were reached.

Several special processes and reports were also provided in sales processing. For example, at the end of the fiscal year, for items that have been back-ordered but are still not received for customers, the system has the ability of cutting a pro forma invoice. These invoices allow the customers to be invoiced before the end of the fiscal year, thereby having their expense accounts charged during the current fiscal year. When the items are received in the subsequent fiscal year, they are handled as having been invoiced. A pro forma status report is provided to all customers to show those items still in a pro forma invoice status. Another special process is the handling of walk-in cus-

TABLE 9.10

```
              CUSTOMER DATABASE  -  ORDER DETAIL                     08/16/89
                                                                     10:26:44
LINE NUMBER    004            DASH NUMBER    01          PP SLIP NO   002
STOCK NUMBER           9156011004
QUANTITY ORDERED            1
QUANTITY SHIPPED            1
UNIT PRICE                 6.73
UNIT COST                  5.8500
PAYMENT NUMBER         T089704
                     ------------ STATUS ------------     --- DATE ---
SHIPMENT               2          SHIPPED                 06/08/89
INVOICING              1          INVOICED                06/08/89
PAYMENT                1          PAID                    06/22/89
DESCRIPTION            57952-005 SPATULA                  SPECIAL ORDER ITEM

OPTION:        TRX:  50D
NEXT LINE-DASH NUMBER:  0012191560110040 1
*** ENTER DATA FOR UPDATE ***
```

tomers. A walk-in customer has an additional markup for our handling of the specific order; therefore, the pricing for walk-in customers is different. Also, sales tax must be charged for walk-in customers. As previously mentioned, backorder relief is automatically performed by the system. Miscellaneous sales reports are provided, as well as various customer cross-reference lists.

9.6 ACCOUNTING INTERFACE

As originally stated, the goals for this system were to provide an easy interface to the then-existing CRS manual accounting books. The original functions as implemented were to generate transactions on a daily or other basis by scanning the customer and vendor databases, to generate daily journals of transactions, and to post these manually to the CRS books. By adding other transactions to the types of transactions generated originally, a full-scale accounting system has subsequently been developed. These transactions go directly into an accounting database, and a full-scale General Ledger Budgetary Accounting System has been implemented that carries on-line detail information about all transactions against specific accounts. The types of transactions handled are:

Journal entries	Invoices
Walk-in Sales	Pro forma invoices
Cash receipts	Monthly billing
Purchase orders	Purchase order adjustments
Receiving reports	Vendor invoices
Completion payments	Retroactive adjustments

Figure 9.6 shows the daily accounting process, which is initiated by scanning databases and generating daily accounting transactions. Journal entries are entered directly through a Journal Entry Screen (Table 9.11), at which time a debit/credit balancing and an account validation function is performed. These journal entries are then added to the daily accounting transactions, produced by scanning the databases, and various accounting reports are provided. Accounting transactions are then stored on tape storage, as well as posted to an accounting database, for subsequent on-line inquiry. The transactions generated by scanning the databases have predetermined general ledger effects. These general ledger effects are easily changed by modifying table entries in the system.

Table 9.12 shows an Account Master (AM) Screen for one account, cash on hand. Amounts are rolled on a monthly basis within the Account Master segment, as well as to the History Segment (IS) (Table 9.13) of the accounting database. Table 9.14 shows a Transaction (TR) Screen for listing current monthly transactions. Individual transactions are kept in this portion of the accounting database, and the amounts are rolled into the appropriate portion

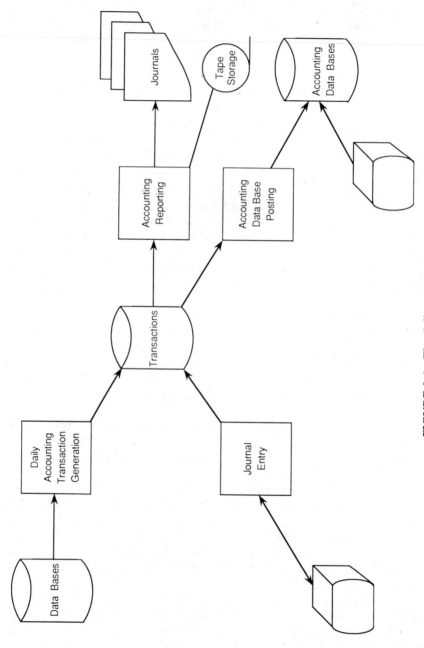

FIGURE 9.6 The daily accounting process.

TABLE 9.11

```
                            C R S   JOURNAL ENTRY SCREEN                 89/08/23
JE NO: 2300          DATE: 890823 (YYMMDD)
REF#1: 05466          REF#2: 6066582     REF#3: VPS#054857
TOTAL DEBIT = CREDIT
ACCT        AMOUNT       D/C    ACCT DESC
114         354.33       D      MERCHANDISE IN TRANSIT
211         354.33       C      ACCOUNTS PAYABLE
422         354.33       D      COST OF MISCELLANEOUS INC
114         354.33       C      MERCHANDISE IN TRANSIT




EXPLANATION:
TO RECORD ORDER AND RECEIPT OF LIQUID NITROGEN.
```

TABLE 9.12

```
VDRS10AC              ACCOUNTING DATABASE INQUIRY              08/16/89
                                                              13:31:22
OPTION:       TRX: 6AM
KEY   : 1111

ACCOUNT NUMBER:  1111
DESCRIPTION:     CASH ON HAND
BUDGT:           .00       ENCUMB:          .00      FREE BAL:    104221.14
                                   BALANCES
                    DEBITS       CREDITS        NET          TYPE
LAST POSTING        29702.74     184163.42      154460.68    C
LAST MONTH          237700.95    168619.41      69081.54     D
MONTH TO DATE       233563.29    189232.44      44330.85     D
LAST YEAR           2591888.21   2601079.46     9191.25      C
YR-TO-MONTH BEG     237700.95    177810.66      59890.29     D
YEAR TO DATE        471264.24    367043.10      104221.14    D
```

TABLE 9.13

```
VDRS10AH                    ACCOUNTING HISTORY                 08/16/89
                                                              13:32:01
OPTION:       TRX: 6IS
KEY   : 1111              1989

ACCOUNT NO:  1989                  YEAR:   1989
DESCRIPTION:  CASH ON HAND
                    DEBITS        CREDITS
                                        STARTING BAL       .00
  MONTH 1   JUL          .00          .00          .00
  MONTH 2   AUG     232647.56    183000.09      49647.47
  MONTH 3   SEP     201501.03    179587.55      71560.95
  MONTH 4   OCT     179943.04    163656.28      87847.71
  MONTH 5   NOV     169974.80    141915.10      115907.41
  MONTH 6   DEC     129092.64    105358.09      139641.96
  MONTH 7   JAN     203950.89    176726.29      166866.56
  MONTH 8   FEB     209195.27    180175.54      195886.29
  MONTH 9   MAR     201183.24    173096.30      223973.23
  MONTH 10  APR     257640.12    240748.49      240864.86
  MONTH 11  MAY     225737.87    202750.65      263852.08
  MONTH 12  JUN     369223.16    712233.20      -79157.96
  CLOSING AMTS     2380089.62   2459247.58         .00
  TOTAL                 .00          .00     ENDING BAL    -79157.96
```

TABLE 9.14

```
          S E C O N D A R Y   K E Y   S E L E C T I O N
RETRIEVE       TRANSACTION: ACCOUNT TRANSACTIONS
OPTION: F   TRX: 6TR   KEY: 1111                    1989
SELECTION:          PRESS ENTER TO VIEW ADDITIONAL SELECTIONS
                                TRANS KEY
    1     8907211114            9202014001C00052                BSC
    2     8907211114            9202014001C00052                BSD
    3     8907211114            9202030001C00052                BSC
    4     8907211114            9202030001C00052                BSD
    5     8907211114            9202030002C00052                BSC
    6     8907211114            9202030002C00052                BSD
    7     8907211114            9202039001C00052                BSC
    8     8907211114            9202039001C00052                BSD
    9     8907211114            9202050001C00052                BSC
   10     8907211114            9202050001C00052                BSD
   11     8907211115            9202014001C00052                BSC
   12     8907211115            9202014001C00052                BSD
   13     8907211115            9202030001C00052                BSC
   14     8907211115            9202030001C00052                BSD
   15     8907211115            9202030002C00052                BSC
   16     8907211115            9202030002C00052                BSD
   17     8907211115            9202039001C00052                BSC
```

of the Account Master Segment on a daily basis. Whenever the current accounting month ends, individual transactions are cleared out and the transaction portion of the accounting database is clear and ready for a new accounting month.

9.7 CONCLUSION

The computer has had a profound impact on all segments of Central Research Stores' operation. It has allowed a rate of change in the data processing area that would not have been possible under a manual or minicomputer system.

The annual operating cost of the computer is somewhat less than the cost of one classified employee. No employees have been added to the operation since the computer arrived, even though we have experienced an annual growth rate of 20% a year. We do not anticipate adding employees in the near future.

The trend for the future seems to be that institutions and corporations are dramatically shifting into computerized operations either with a minicomputer or hooked to a mainframe. It must be emphasized that no system should be implemented without proper safeguards for security, recoverability, and auditability. The 1990s will certainly be challenging if the rate of change in the data processing area continues to grow at the rate at which it has over the past decade. It becomes almost inconceivable to visualize what administrative roles will be like for direct users and data processing professionals in the year 2000.

CHAPTER 10

SURVEYS AND INSPECTIONS AT ACADEMIC CHEMICAL STORAGE FACILITIES

PATRICIA ANN REDDEN
Professor of Chemistry
St. Peters College
Jersey City, New Jersey

10.1 INTRODUCTION

Many conferences, short courses, and symposia at professional meetings are devoted to solving problems involved in storing and handling chemicals. Equally active are governmental agencies such as the National Institute for Occupational Safety and Health (NIOSH), the Occupational Safety and Health Administration (OSHA), the Environmental Protection Agency (EPA), and the U.S. Consumer Product Safety Commission (CPSC), and professional groups such as the American Chemical Society (ACS) and the National Science Teachers' Association (NSTA). Books and journal articles proliferate on the question of safety in working with chemicals.

This chapter identifies the current conditions of laboratory safety in a select group of the academic community. Both industrial and academic facilities are subject to the safety regulations of governmental agencies, which conduct regular inspections to ensure the compliance that results in worker safety. However, such inspections are rare in academic institutions. In most cases, academic institutions must address the safety issue without the guidance provided by these inspections. Unsafe practices in academic laboratories result in special problems that must be recognized and corrected.

10.2 SOURCES OF DATA

The main sources of our data will be the reports of the Safety Committee of the New York Section of the American Chemical Society, surveys taken by Lab Safety Supply Company, a memorandum prepared by the Chemical Hazards

Program of the U.S. Consumer Product Safety Commission, and a published survey on the results of OSHA-type inspections of academic institutions. The data from these surveys will give the reader a better understanding of the scope of these problems.

The Safety Committee of the New York Section of the ACS (hereafter referred to as N.Y.–ACS Safety Committee) was first formed in 1979. There are currently 8 members from academic institutions and one nonacademic member. Each year a letter is sent to the colleges in the New York Section inviting them to take advantage of an on-site inspection. The standards used for these inspections are described in the American Chemical Society's publication, *Safety in Academic Chemistry Laboratories* (1). The method and forms used in the inspections have been previously discussed in detail (2), as have the results of the first year's inspections (3).

During the spring of 1980, the Safety Committee conducted on-site inspections of 11 colleges in the New York metropolitan area. In 1981 10 of these colleges and universities were reinspected, and eight additional institutions were inspected for the first time. Two more colleges were inspected in 1982. The 21 inspected colleges are in an area that includes New York City, Westchester, part of northern New Jersey, and all of Long Island. Nine public colleges and 12 private colleges comprise the list. Four have college graduate programs, 15 grant only the baccalaureate degree, and 2 are community colleges. The number of full-time faculty in the chemistry departments generally ranges from 3 to 12, although institutions with graduate programs have considerably larger faculties. The number of students using the chemistry laboratories per week varies between 100 and 2000. The number of chemistry majors is reported as between 10 and 150.

Each institution* was asked to complete a preinspection form. Then a team that consisted of two to six inspectors, depending on the size of the college and the number of laboratories, conducted the inspection. The results were summarized and sent to the department's safety officer or chairperson with recommendations for improvement where needed. The same procedure was followed for reinspections. Careful note was taken of previously made recommendations and attempts made to implement them. The inspection committees were so thorough that, to the surprise of many of the inspected institutions, storage areas and laboratories were inspected bottle by bottle and bench by bench. Since most of the inspectors were academicians, they were particularly aware of the hazards posed by relatively inexperienced students handling chemicals and equipment. At the same time, the constraints (financial and otherwise) that handicap chemistry departments trying to make their facilities safer were clearly understood. The recommendations of the committee were made with these points in mind.

---

*Since the intention of this chapter is not to present the problems of specific colleges, but to draw conclusions about college laboratory safety that may be informative to the scientific community as a whole, the names of the colleges inspected are withheld.

Lab Safety Supply Company personnel conducted surveys at the Pittsburgh Conference (March 1981 in Atlantic City, NJ) and at the National Science Teachers' Association's National Convention (April 1981 in New York City). Excerpts from these surveys were published in the *Journal of Chemical Education* (4). One hundred sixty-five questionnaires were completed at the Pittsburgh Conference. Unfortunately, there is no breakdown of the results to indicate how many of the respondents were from industry and how many were from academic institutions. The questionnaire at the NSTA convention was completed by 143 science educators, the majority of whom taught in grades 7–12. A breakdown by grade level and discipline taught is found in Table 10.1. The questionnaires themselves and the responses are given in Tables 10.2 and 10.3.

The U.S. Consumer Product Safety Commission (CPSC) conducted a limited survey through its 11 regional and district offices. Each office was requested to contact the safety officers or administrators of two school districts (one rural and one urban). Each administrator was asked to provide information on the safety equipment available in the school district, describe the methods of chemical disposal used, and list the chemicals used in the laboratory. The information was presented to the commission in a memorandum prepared by Rory Sean Fausett, Program Manager of the Chemical Hazards Program, H.S., in January 1982(5). The memorandum reports the results of a search of database files for reports of school-related injury due to the use of chemicals in secondary schools. The files studied were the National Electronic Injury Surveillance system (NEISS), accident investigation reports, injury or potential injury reports (IPII), death certificates, and a National

TABLE 10.1 Distribution of Survey Responses at NSTA Conference, April 1981

	Number of Responses	Percentage of Total Responses
Breakdown of Responses by Teaching Level		
K–12	6	4
K–6	8	6
7–12	115	80
College	10	7
No response	4	3
	143	
Breakdown of Responses by Subject Matter Taught		
General science	34	24
Biology	30	21
Chemistry	67	47
Other	12	8
	143	

TABLE 10.2   Survey Conducted at NSTA Conference, April 1981

Question	Yes	No
Do you have a chemical storeroom?	90%	9%
Is your storeroom normally left unlocked?	18%	79%
Does your storeroom have two or more clearly marked exits?	43%	52%
Are the aisles in your storage area free from obstruction?	66%	30%
Is the air in your storeroom dehumidified?	13%	82%
Are the shelves on which you store chemicals fastened to the wall or floor?	73%	20%
Do the bottles in which you store chemicals have labels with safety and/or first aid information?	46%	51%
Are bottles of chemicals labeled with receiving and expiration dates?	28%	69%
Are chemicals arranged alphabetically on the shelves?	73%	25%
Are chemicals arranged by class (oxidizers with oxidizers, flammables with flammables, etc.)?	47%	51%
Are chemicals stored next to a heat register, radiator, or other heat source?	4%	96%
Are stored chemicals exposed to direct sunlight?	5%	94%
Do you have a special cabinet for storing flammable liquids?	56%	41%
Do you use safety cans while storing flammables on the laboratory bench or counter?	24%	70%
Are large bottles of concentrated acids or bases stored above waist level?	13%	86%
Are bottles of concentrated acids stored separately from inorganic bases?	76%	24%
Do you use bottle carriers for transporting acid bottles?	11%	87%
Do you use neutralizing agents for spills of acids or bases?	64%	32%
Do you use absorbents for cleaning up chemical spills?	35%	61%

Source: Courtesy of Lab Safety Supply Company, Janesville, WI.

Safety Council publication entitled *Accident Facts.* This does not give a full picture of accidents resulting in injuries, since only injuries severe enough to warrant hospital emergency room treatment were reported, and the NEISS coding system did not allow a national projection to be made. However, the data are of interest when combined with information about chemical handling practices in the schools.

A final source of information is a 1977 report by Schmidt (6) on the results of OSHA-type inspections of academic institutions. At that time, 19 institutions responded to a survey of chemistry department chairpersons taken throughout the United States, indicating that they had undergone an OSHA-type inspection. Nine of these inspections were carried out by state and federal

OSHA, four by campus health and safety offices, one by a state department of labor, two by an insurance company, two by a consulting firm, and one by a visiting committee. Twelve Ph.D.-granting institutions were inspected as compared to seven undergraduate departments.

10.3 DESCRIPTION OF PHYSICAL FACILITIES

Many laboratories were built before the issue of safety was so much in the forefront as it is today. As a result, necessary safety equipment is either missing or retrofitted after construction of the laboratory. The purchase of this equipment can be a real strain on a department budget. Proper installation, maintenance, and use requires both concern for and knowledge of laboratory safety principles. Discrepancies can easily occur between what an institution or instructor believes is adequate and what is recommended by agencies knowledgeable in safety. Pertinent to physical facilities are the topics of ventilation, chemical storage facilities, personal and site safety equipment, communications and egress, and housekeeping.

10.3.1 Ventilation

Ventilation is frequently a major problem in chemical storage areas, although laboratories are normally adequately ventilated. This is particularly true of those storage areas in academic institutions that are located in a basement or a converted, often windowless cubbyhole. The ventilation in those storerooms often was not safe for the quantities of volatile chemicals that were stored. Of the 21 colleges inspected by the N.Y.–ACS Safety Committee, 11 were found to have this type of problem. In one case, the ventilation was so poor that the inspectors developed severe headaches after relatively short exposure to storeroom atmosphere. The survey taken at the Pittsburgh Conference indicates that only 44% of those responding had chemical storerooms that exhausted air out of the building, which would reduce the severity of ventilation problems.

   Fume hoods in laboratories, on the whole, performed poorly in the on-site inspections. Less than half of the colleges had hoods that were all operable and working at the recommended velocity (100 cubic feet per minute), and only one college had marked the sash level at which this velocity was achieved. In fact, in several institutions the hood sashes had to be almost closed to operate correctly. Some institutions had inoperable sashes on hoods, and many had interior controls for gas, water, and so on. Adequate lighting in hoods was a problem in many cases. In one college the hood sides and sash were made of ordinary window glass rather than safety glass or wire-impregnated glass.

   It is of interest to note that a list of safety problems found in the OSHA-type inspections of academic laboratories (6) listed inadequate ventilation and exhaust hood flow velocity as the fourth most frequent problem (Table 10.4).

TABLE 10.3 Survey Conducted at Pittsburgh Conference, March 1981

Question	Yes	No
1. Are chemicals stored in a specially designated storeroom? (if not, go on to question 4)	67%	32%
2. The chemical storeroom		
a. Is locked at all times, with entry by authorized personnel only	36%	30%
b. Is identified with a sign as a chemical storeroom	42%	30%
c. Has two or more clearly marked exits	33%	30%
3. The chemical storeroom		
a. Has sufficient lighting to read labels of chemicals	65%	6%
b. Has a ventilation system that exhausts room air to the outside of the building	44%	23%
c. Has a cool, dry atmosphere (either air conditioning or dehumidifier system)	46%	22%
d. Has unobstructed aisles (i.e., no blind alleys)	48%	17%
e. Has shelving units attached to the wall or floor	52%	14%
4. All chemical containers are clearly labeled		
a. As to their contents	90%	5%
b. With receiving and disposal dates	32%	58%
5. Equipment and procedures readily available for emergencies include		
a. Approved first aid supplies	72%	20%
b. Posted emergency telephone numbers	64%	28%
c. Eyewash facilities	83%	13%
d. Drench shower facilities	81%	14%
e. Spill cleanup supplies	60%	33%
f. Fire extinguishers	92%	3%
g. Self-contained breathing apparatus	48%	44%
6. Chemicals are stored		
a. Alphabetically	63%	31%
b. By class—oxidizers with oxidizers, flammables with flammables, etc.	45%	39%
c. By random placement	23%	59%
d. In a cool, dry atmosphere	66%	25%
e. Away from direct sunlight or localized heat	87%	7%
7. Chemicals are stored in the following manner		
a. Incompatible chemicals are physically segregated	65%	24%
b. Large bottles of acids are stored on a low shelf or in acid cabinets	82%	12%
c. Oxidizing acids are segregated from organic acids and flammable and combustible materials	70%	20%
d. Acids are separated from inorganic bases	70%	21%
e. Acids are separated from active metals such as sodium, potassium, magnesium, etc.	81%	10%
f. Acids are separated from chemicals that could generate toxic gases on contact: iron sulfide, sodium cyanide, etc.	79%	12%

TABLE 10.3   (*Continued*)

Question	Yes	No
g. Flammable liquids amounting to more than 1 pint are stored in approved safety cans or cabinets	61%	32%
h. Peroxide-forming chemicals (e.g., diethyl ether) are stored in airtight containers in a dark, cool, dry place	72%	18%
i. Peroxide-forming chemicals are labeled with receiving, opening, and disposal dats	39%	46%
j. Water-reactive chemicals such as calcium oxide are stored in a cool, dry place away from any potential water source	65%	24%
k. Oxidizers are physically segregated from flammable and combustible chemicals or materials	68%	22%
l. Oxiders are physically segregated from reducing agents such as zinc, alkaline metals, and formic acid	64%	22%

Source: Courtesy of Lab Safety Supply Company, Janesville, WI.

This seems to correlate with the findings of the N.Y.–ACS Safety Committee. It is strongly suggested that each institution purchase an inexpensive vane anemometer to check the hoods for proper operation and then mark the optimum sash height for future references.

10.3.2 Chemical Storage—Stockrooms

Although all the colleges inspected by the N.Y–ACS Safety Committee and 90% of the educators surveyed at the NSTA conference had separate chemical storerooms, approximately one-third of the responses at the Pittsburgh Conference indicated that chemicals were not stored in a specially designated storeroom. If separate storerooms existed, approximately half did not have two or more clearly marked exits. The inspected colleges almost invariably kept the storerooms locked, with entry by authorized personnel only, but only 79% of the responses at the NSTA conference and 50% of those at the Pittsburgh Conference with separate storerooms had a similar arrangement. Of the responses at the Pittsburgh Conference, 50% indicated that the chemical storeroom was not clearly identified.

Most of the colleges inspected by the N.Y.–ACS Safety Committee did not have adequate storage facilities for volatile or flammable chemicals. Some had "solvent rooms" with quite inadequate facilities. These were usually ordinary storage rooms filled with volatile or flammable materials. Although some colleges employed below-ground or outside bunkers, there were no provisions for maintaining a moderate temperature to minimize summer heat or winter freezing. The majority of colleges inspected, despite need, were unable to purchase needed solvent storage cabinets because of the fairly high cost.

TABLE 10.4   Safety Problems in Academic Laboratories[a]

Improper electrical wiring
  Ungrounded equipment
  Overloaded circuits
  Inappropriate high voltage shielding
Unguarded belt and pulley asemblies, saw blades, and buffer wheels
Improper storage of bulk chemicals
  Stockroom
    Design of shelves, air-handling system, fire equipment
    Ungrounded bulk solvent drums
  Laboratories—research and instructional
    Excessive volumes of solvent
    Lack of metal safety cans and metal cabinets
    Lack of explosionproof refrigerators
  Carcinogenic compounds
Inadequate ventilation and exhaust hood flow velocity
Lack of eyewash and safety shower facilities
Less major problem areas
  Waste disposal
  Storage and use of gas cylinders
  Poor housekeeping practices
  Inadequate safety signs
  Inadequate shields for individual experiments
  Blockage of escape routes between research areas

[a] Identified by OSHA-type inspections (in approximate order of the frequency by citation)(6).

Source: Reprinted with permission from R. L. Schmidt (1977). "Academic Experiences with O.S.H.A.," *J. Chem. Ed.,* **54**, A145.

Two colleges built wooden storage cabinets conforming to OSHA standards (7,8), at a greatly reduced cost.

In the absence of suitable solvent cabinets, either fume hoods or open shelving was used for storage. Occasionally the solvents were stored in separate wooden or steel cabinets that did not conform to the safety standards. This situation was also verified by responses at the NSTA conference. Only slightly more than half of those surveyed indicated the presence of special cabinets for storing flammable liquids.

Areas in which solvents are dispensed should have facilities to ground and bond solvent cans and drums. Only three inspected colleges had such arrangements. This was also cited as a recurring problem in the report of the OSHA-type inspections.

Although large temperature fluctuations are undesirable in chemical storage areas, discussions with high school teachers indicate this to be a serious problem. Since high school laboratories seldom are used through the summer, the rooms are unventilated and may become extremely warm. One high school instructor in New Jersey reported to the author that a block of paraffin stored on an open storeroom shelf was discovered melted when class resumed in September!

A final note should be made on acid storage cabinets and explosion-proof refrigerators. Few inspected colleges had adequate storage for acids. Most colleges stored acids on open shelves in a general storage area or in cabinets under hoods. Segregation of acids from incompatible chemicals was generally uncertain. Explosion-proof refrigerators full of nonvolatile and nonexplosive chemicals indicated uncertainty about their proper usage.

10.3.3 Personal and Site Safety Equipment

Personal and site safety equipment are frequently deficient in academic laboratories. Each laboratory should contain an adequate number of fire extinguishers of appropriate type and size, a safety shower, piped-in eyewash unit(s), spill control materials, and a minimum amount of first aid supplies, particularly sterile pads and bandages. Additionally, chemical storage areas should be equipped with automatic fire extinguishers and fire and/or smoke detectors. Self-contained breathing apparatuses and materials for control of larger spills must be readily available for emergencies. Unfortunately, these supplies are often missing, inadequate, or poorly sited.

Eyewash units in particular are a problem. Several colleges and high schools depended solely on small, portable units containing approximately 500 mL of solution. Since the ACS recommendation is that the eyes be flushed for 15 minutes (1), these units are woefully inadequate and should be used only to obtain access to a piped-in eyewash. In the inspected colleges, piped-in eyewash units were mounted on the cold-water line rather than supplying tepid water. Many eyewash units required the use of one hand to keep the water running, making it difficult to hold the eye open. Other units were placed so low that students would have to stoop to wash their eyes. Additionally, an inadequate number of eyewash stations often existed, so a student might have to run across an entire laboratory or half-way down a hall to use the eyewash. The lack of proper eyewash facilities is cited in the report on OSHA-type inspections and appears in the CPSC report as well.

Safety showers also have their share of problems. Every college inspected by the N.Y.-ACS Safety Committee had emergency safety showers, but these were not always adequately located or engineered. One college had a valve instead of a pull ring on the shower, which reduced the availability and the rate of delivery of the water. Another college had rings that were held over door-ways by cup hooks, necessitating an upward movement before a downward pull. (Interestingly, this arrangement was approved by an industrial safety consultant, the local fire department, and the insurance company's inspectors. Only the academic members of the N.Y.-ACS Safety Committee pointed out that this would be a real hazard for students inexperienced in an emergency.) Other colleges had showers that were jointly used by two or more instructional laboratories, rather than one shower per laboratory. Even when showers were adequate, there was only one college that indicated when the showers had last been checked; most were not checked at all. Only six of the school districts responding to the CPSC survey said that they had safety showers, and this lack was also cited for the OSHA-type inspections.

The American Chemical Society recommends that each laboratory bench be equipped with one fire extinguisher of appropriate size and type (1). This might seem somewhat excessive to many chemists, and in actuality the New York–ACS Safety Committee evaluated the number of extinguishers on the basis of room size, geometry, and extinguisher type and size. Despite this reduced requirement, few inspected laboratories had adequate coverage. These deficiencies were particularly noticeable in storage areas, even though it is these areas that present the greatest potential hazard. Many extinguishers in laboratory areas were poorly marked and not mounted accessibly. Others lacked seals or were partially discharged, making them undependable in an emergency. Many sealed extinguishers did not carry tags with current dates of inspection.

Laboratory and storage areas often have no automatic alarm and detection system for smoke and/or heat. If such a system is present, it frequently rings only within the department or the building, not at a central security desk or local fire department. Isolated chemical storage areas in particular should have not only a detection–alarm system but adequate automatic fire extinguishers as well.

Figure 10.1 illustrates the type of poor laboratory planning that was found in one institution inspected by the N.Y.–ACS Safety Committee. The placement of the combination safety shower–eyewash unit at the back of the laboratory between two hoods limits access by those students working at distant tables. The constricted space also makes it difficult for an instructor to administer aid to an injured student using this equipment. The college would be better off with two eyewash units mounted on the sinks of alternate laboratory tables, at the end opposite the hoods. The partially discharged fire extinguishers shown in Figure 10.1 were poorly sited for maximum accessibility.

Every laboratory should have a policy requiring the use of safety goggles or glasses. Most states mandate such protection in academic laboratories. However, when touring academic laboratories, one will invariably discover some students and faculty members working in the laboratory without proper eyewear. The use of prescription glasses without eye shields is often allowed, and the ACS warning against contact lenses (1) is usually acknowledged but rarely enforced. Many faculty members report resistance to the use of goggles by their students, primarily because of poor fit or scratched lenses that result in a visibility problem.

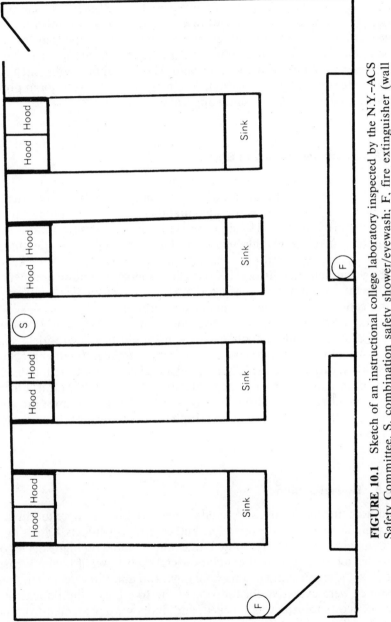

FIGURE 10.1 Sketch of an instructional college laboratory inspected by the N.Y.–ACS Safety Committee. S, combination safety shower/eyewash; F, fire extinguisher (wall mounted).

Only about one-third of the colleges inspected by the N.Y.–ACS Safety Committee had adequate emergency breathing apparatus available. This apparatus was not always stored in an accessible, well-marked location. The apparatuses found most often were emergency oxygen bottles and filter masks. Only three colleges had self-contained breathing apparatuses located outside the chemical storerooms. The predictable location for emergency breathing apparatus was in the storeroom. The danger of keeping such equipment in the storeroom occurs when an emergency in the storeroom prevents entry without a proper breathing apparatus.

10.3.4 Communication and Egress

In case of an emergency in a storage area (especially in a basement or bunker) or in a laboratory, communication with the outside is imperative. In many cases the college inspections revealed that there were no telephones nearby. This problem can be solved by simply installing an in-house telephone line or an intercom, in the event that a school is reluctant to have telephones that may be used by unauthorized persons. This phone should have a sticker or a nearby prominent sign listing the following emergency phone numbers: fire department, hospital emergency room, on-premises medical staff, poison control center, and emergency chemical information.

Emergency exits from laboratories and storage areas should be clearly marked and accessible. Ideally each room should have two widely spaced exits. Inspection revealed that the second exit was frequently either blocked or locked for security reasons. Where a second exit does not exist, consider a clearly marked emergency escape ladder or kick-out panel. At least one New Jersey high school painted bright yellow paths to the nearest egress directly on the floor, so that students could exit in the correct direction from a smoke-filled room.

10.3.5 Housekeeping

Conditions in storage areas were not very good. Shelves were too high and aisles too narrow for safely removing and carrying chemicals. Labels on bottles were missing or deteriorating. Large bottles stored on high shelves often protruded over the edges of the shelves. Metal shelves were frequently rusting. Bottles, solvent cans, and drums were piled two and three deep. These large quantities were excessive for the size of the room, and the floor area was covered, access to exits was blocked, and doors were prevented from being opened or closed. Finally, spill control was a problem with poorly labeled and hard-to-find neutralizing and containment materials.

These problems also appear in the report on the OSHA-type inspections of academic institutions.

10.4 CHEMICAL STORAGE AND DISPOSAL

10.4.1 Inventory Control

As student enrollments in laboratory courses drop, faculty turnovers occur, and laboratory syllabi change, the need for chemicals varies. However, ordering practices do not always reflect these changes. Thus large amounts of chemicals accumulate in quantities that cannot be used even over several years. In many cases these chemicals are hazardous or possess a limited shelf life. One college inspected by the N.Y.–ACS Safety Committee had a 65-lb container of ammonium nitrate in a solvent storage room. Another had over 50 one-gallon bottles of glacial acetic acid on its shelves. Large quantities of flammable solvents were found in several colleges. Diethyl ether, known to form explosive peroxides, was often undated or out of date. Another college had been "blessed" with a donation of fuming acids, alkali metals, and lachrymators some years ago and was desperately trying to store and dispose of them properly. Economics has been a reason for ordering large quantities of chemicals in that a better unit price can be obtained. Four of the school districts in the CPSC survey indicated that chemicals were ordered in large volume to cut costs. Sixteen of the school districts store chemicals from year to year rather than order only what will be used for the current school year.

Proper inventory control can often minimize the excesses. A system of dating materials on arrival encourages the use of older chemicals first. Only 28% of those completing surveys at the NSTA conference and 32% of those at the Pittsburgh Conference indicated that chemical containers were clearly marked with receiving and disposal dates. This is of particular importance for ethers and peroxide-forming chemicals, which should also be marked with the date they were opened (9).

10.4.2 Compatibility Storage

Compatibility storage costs nothing but time. The N.Y.–ACS Safety Committee found that only a few of the 21 inspected colleges attempted to separate stored chemicals according to their reactivities. In only three cases was the segregation respectable. Most storage rooms segregated only strong acids from concentrated bases or kept organic compounds separate from inorganics. Flammable or volatile organic solvents, unless in drums or cans, were usually arranged alphabetically among other organic compounds. Bromine and sodium metal were often stored on common shelves. Bromine and nitric acid, both strong oxidizing agents, were frequently in close proximity to organic chemicals. Concentrated hydrochloric acid and ammonium hydroxide were found side by side. The survey responses in Tables 10.2 and 10.3 indicate the universality of this problem.

Figure 10.2 illustrates the magnitude of the problem when many of these individual hazards are combined in a chemical storeroom. This is a sketch of a

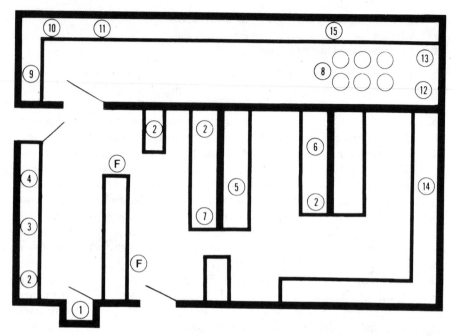

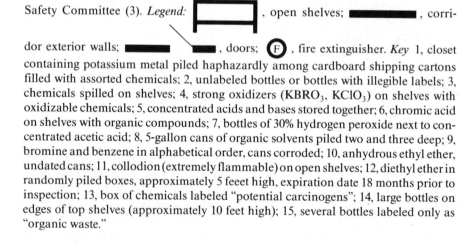

FIGURE 10.2   A chemical storage area of a college inspected by the New York–ACS Safety Committee (3). *Legend:* [ ] , open shelves; ▬▬▬ , corridor exterior walls; ▬▬▬ , doors; (F) , fire extinguisher. *Key* 1, closet containing potassium metal piled haphazardly among cardboard shipping cartons filled with assorted chemicals; 2, unlabeled bottles or bottles with illegible labels; 3, chemicals spilled on shelves; 4, strong oxidizers ($KBRO_3$, $KClO_3$) on shelves with oxidizable chemicals; 5, concentrated acids and bases stored together; 6, chromic acid on shelves with organic compounds; 7, bottles of 30% hydrogen peroxide next to concentrated acetic acid; 8, 5-gallon cans of organic solvents piled two and three deep; 9, bromine and benzene in alphabetical order, cans corroded; 10, anhydrous ethyl ether, undated cans; 11, collodion (extremely flammable) on open shelves; 12, diethyl ether in randomly piled boxes, approximately 5 feeet high, expiration date 18 months prior to inspection; 13, box of chemicals labeled "potential carcinogens"; 14, large bottles on edges of top shelves (approximately 10 feet high); 15, several bottles labeled only as "organic waste."

storage area found in one of the colleges inspected by the N.Y.–ACS Safety Committee. Trouble spots are indicated.

There is no good reason why storage of chemicals according to compatibility cannot be achieved. The most common reason given for not storing chemicals according to compatibility deals with the lack of compatibility information. However, there is a list of incompatible chemical combinations in the appendix of the ACS manual (1). Pipitone (10) has created a checklist that can be used as a starting point to achieve good storage practice. In addition, the manufacturers' labels on chemical containers give a good deal of information about reactivities and special storage needs (11). Another good source for information on both correct storage and disposal methods is the chemical catalog reference manual of Flinn Scientific, Inc. (12). Many articles and books have appeared describing inventory methods and labeling techniques to convey information about safety, storage, and disposal (4,13–15).

10.4.3   Toxic Substances and Carcinogens

In its survey, the CPSC requested a list of chemicals used in the 22 secondary school districts. The total list included 312 chemicals. No attempt was made to differentiate between those actually in use and those kept in stock but not used. The hazardous properties (if any) of these chemicals were assessed using *Dangerous Properties of Industrial Materials,* 5th ed., by N. Irving Sax (16); System for Tracking the Inventory of Chemicals (STIC); the *NIOSH/OSHA Pocket Guide to Chemical Hazards* (15); and a commercial chart, "Toxic and Hazardous Chemicals in Industry," prepared by Science Related Materials, Inc. (17). The STIC is a new CPSC database using listings of carcinogens by OSHA, the EPA's Carcinogen Assessment Group (CAG), and the Department of Health and Human Services' (DHHS) first *Annual Report on Carcinogens.*

Table 10.5 lists those chemicals in the schools' inventories classified by Sax or the STIC database as potential or recognized carcinogens. There is no universal agreement on these classifications. Sax identifies 28 potential carcinogens and OSHA only four. Thirteen of those listed by Sax have been reviewed by the International Agency for Research on Cancer (IARC). Of these, seven were classified as carcinogens and two as suspected carcinogens. For the remaining four there was insufficient data to make a proper classification. Eleven of the chemicals were identified as potential teratogens, whereas two are possibly mutagenic. It is interesting that benzene is on the list. Many older laboratory manuals ascribed benzene as a standard solvent. Newer editions of the manuals reflect a change to other solvents. However, high schools often use the same editions for several years after publication since the books are purchased by the school.

The presence of potential carcinogenic chemicals in the surveyed secondary school districts is evident from the findings of the OSHA-type inspections (see Table 10.4) and from the inspections by the N.Y.–ACS Safety Committee (cf. item 13 in Figure 10.2). Many institutions are aware of this problem and are taking steps to either eliminate or safely identify and store carcinogens. Others indeed seem oblivious to the issue.

Carcinogenicity is not, of course, the only hazard identified with chemicals. Compounds such as linseed and cottonseed oils (allergens), hydrosine sulfate (a poisonous alkaloid), and lithium hydroxide (very caustic and toxic) are available to students, often with insufficient warning. The "Toxic and Hazardous Chemicals" safety chart used by CPSC rates chemicals according to toxicity or health hazard, flammability, and chemical reactivity on a scale of 0–4.

TABLE 10.5 Compounds Identified by the 1982 U.S. CPSC Report as Carcinogenic, Teratogenic, Mutagenic, or Causing Neoplasm Formation (5)[a]

| | Identification by Sax (16) | | | | Identification by STIC | | |
	Carcinogen	Teratogen	Causes Neoplasms	Mutagen	DHHS	EPA-CAG	OSHA
Acetamide	exp (+)						X
Aniline hydrochloride	exp (±)						X
Benzene	Recognized leukemogen				X	X	X
Benzidine	recog			X			
Cadmium chloride	exp (+)	exp	exp		X	X	
Cadmium nitrate					X	X	
Carbon tetrachloride	exp (+)	exp	exp			X	
Chloroform	exp (S)		exp (S)			X	
Chromic acetate	recog		recog				
Chromic acid	exp (+)						
Colchicine	exp	exp	exp	exp			
Dichloromethane	exp						
1,2-Dichlorobenzene	exp (±)						
2,4-Dichlorophenol	exp						
Diphenylamine	exp	exp					
Ethylene dichloride	exp	exp		exp			
Ferric oxide	exp (±)						
Formaldehyde	exp					X	

Substance						
Isoamyl alcohol	exp					
Isobutyl alcohol	exp					
Kerosene	susp		exp			
Lauric acid	exp (+)					
Lead acetate		exp				
Lead chloride		exp				
Lead nitrate		exp				
Lithium chloride		exp				
Methyl ethyl ketone		exp				X
Nickel(ous) ammonium sulfate					X	X
Nickel(ous) chloride					X	X
Nickel(ous) sulfate			exp		X	X
Phenol	exp					
1-Propyl alcohol	exp					
Pyrogallic acid	exp					
Salicylamide		exp		poss		
Sodium chromate	exp (±)					
Sodium dichromate	exp (S)					
Tannic acid	exp (+)					
Thioacetamide	exp (+)					X

[a] *Symbols used:* exp, experimental; poss, possible; recogn, recognized; susp, suspected; (+), on review by International Agency for Research on Cancer (IARC), classified as a carcinogen; (S), on review, IARC classified it as a suspected carcinogen; (±), on review, IARC states there is insufficient data for classification; X, identified by the group named as a carcinogen (potential or recognized).

A health hazard rating of three means that short-term exposure may result in major temporary of permanent injury and may threaten life. A rating of 4 means that short-term exposure may result in major permanent injury or death. In the CPSC survey, 31 chemicals had a health hazard rating of 3 and two had a rating of 4. Recognition of the hazard and careful usage minimize the risks for many of these compounds, whereas some should definitely be eliminated from an academic institution's inventory.

A *Manual of Safety and Health Hazards in the School Science Laboratory* (18) has recently been published by NIOSH. This manual evaluates the hazards connected with chemicals, equipment, and procedures in experiments typical of high school courses in chemistry, biology, earth science, and physics. Each chemical is assigned a health rating as those on the previously mentioned safety chart. In addition, an appendix identifies chemicals with greater hazardous nature than potential usefulness, chemicals that should be removed from the schools if alternatives can be used, or chemicals that should be retained only in minimal amounts. Detailed information is given on the toxicity, dose levels, and routes of exposure of carcinogens (positive, potential, or suspected) and of other toxic compounds.

10.4.4 Spill Control

The topic of spill control must be addressed on both large and small scale, for spills of both solids and liquids (4).

Solid materials may be cleaned up fairly easy with a dustpan and brush, but their treatment after cleanup should be carefully considered. Dumping a strong oxidant such as potassium perchlorate in a wastebasket full of paper is certainly not a recommended procedure as fires can occur! Compatibility should be considered when several chemicals are being placed into the same waste container. The shelves in several of the storage areas inspected by the N.Y.–ACS Safety Committee contained spilled chemicals that had not been cleaned. Many storage areas did not have containers for collection of solid waste.

Liquid spills are a more treacherous problem than solid spills because of seepage into cracks, flow properties that can carry the material under doors and cabinets, and volatility. The liquid spilled must often be neutralized as well as removed. Traditional methods of controlling liquid spills include the use of sodium bicarbonate to neutralize acid spills and diatomaceous earth, sand, or vermiculite to absorb organic materials. Currently available commercial products include pillows of various sizes filled with an absorbent that can absorb up to 98% of the pillow's capacity in less than 30 seconds. The whole pillow, with the absorbed contents, can then be disposed of correctly and safely. Deep trays under bottles and sills in solvent storage rooms to contain spills are also recommended.

The survey at the NSTA conference showed that 64% of the respondents used neutralizing agents for acid or base spills, but only 35% used other absorbents. At the Pittsburgh Conference, 60% of those surveyed used spill cleanup supplies (undifferentiated by type). The N.Y.–ACS inspections also identified a lack of adequate materials in many of the inspected colleges.

A final point should be raised on this topic. Chemical spills may cause moderate to severe burns to the person coming into contact with the material. Neutralizing materials such as sodium bicarbonate paste or solution (for acid burns) and sodium thiosulfate solution (for bromine spills and burns) should be placed adjacent to the area where these chemicals are to be used and should be prominently labeled for emergency first aid. This is particularly important when a minor spill down the side of a reagent bottle may be temporarily undetected until an accident occurs.

10.4.5 Waste Disposal

Disposal of waste chemicals is a major problem confronting chemical laboratories, particularly academic laboratories. Each college inspected by the N.Y.–ACS Safety Committee was in the process of developing its own method for the safe disposal of chemical wastes. This leads, however, to the practice of keeping unlabeled bottles and waste solvents in storage areas since safe, legal disposal techniques are currently too expensive for many high schools and colleges. Some colleges have found that their local fire departments or bomb squads are willing to help remove their waste. Another college recovers some of its solvents by fractional distillation. Unfortunately, two of the colleges surveyed have been dumping the waste down the drain and depending on acid traps to neutralize the liquids. Acid traps have no effect on the organic solvents that are disposed of in this manner.

The CPSC survey of chemicals in secondary school laboratories reveals that 20 of these chemicals are classified as hazardous wastes by the EPA's Resource Conservation and Recovery Act (RCRA). A listing of these chemicals can be found in Table 10.6. Table 10.4 indicates that proper disposal is a problem in the institutions that underwent OSHA-type inspections.

10.5 CONCLUSION AND RECOMMENDATIONS

This chapter has emphasized problem areas from the data in surveys and inspections in laboratories and storage areas. There are many such problems. However, it would be remiss of the author not to point out the positive signs as well. The inspections of the N.Y.–ACS Safety Committee and the surveys referred to in this chapter indicate that many academic institutions and chemical companies are making a serious effort to improve their safety posture and are indeed succeeding. The participation of academic institutions in voluntary inspection programs is an obvious example of this trend, as is the attendance of faculty members and chemists at safety workshops and symposia. The continuation and expansion of these programs is essential.

TABLE 10.6   Compounds Identified on U.S.CPSC Survey of Chemicals in Secondary Schools That Are on the RCRA List

Acetone	Lactic acid
1-Butanol	Methyl methacrylate
Benzene	Naphthalene
Benzidine reagent	Nickel(ous) ammonium sulfate
Cyclohexane	Nickel(ous) chloride
Carbon tetrachloride	Nickel(ous) sulfate
Chloroform	Phthalic anhydride
Dichloromethane	Thioacetamide
Formic acid	Toluene
Isobutyl alcohol	Xylene

Inspection of academic laboratories and storage areas, by an external professional group (as in the case of the N.Y.–ACS Safety Committee) with a knowledge of institutional needs and problems, can point to the ways and means of improvement. Most of the colleges inspected used the committee's reports as support for financial requests by the department for needed safety equipment. Problems that had been overlooked even by hard-working safety officers and committees were revealed by the inspectors' visit. As a result of the first year's inspection reports, all but one college showed substantial improvement after reinspection one year later. The greatest improvements were made in the area of safety equipment and housekeeping. Piped-in eyewash units, proper safety showers, adequate fire extinguishers, and safety signs were relatively easy additions to the laboratory. Housekeeping problems were also easily remedied. In contrast, neither chemical storage nor structural problems were addressed in a majority of cases. The major reason is the lead time needed to complete the work. The acquisition of funds was not possible in such a short time period.

ACKNOWLEDGMENTS

The author thanks the New York Section of the American Chemical Society and the society's Program Development Fund for their help, both financial and moral, in setting up the Safety Committee and supporting its work. Gratitude is expressed to the safety officers and faculty members of the inspected colleges for agreeing to the publication of our results. Finally, gratitude is due to the members of the Safety Committee who have donated many hours of time, both on inspections and in preparing the following reports to the colleges.

The author also wishes to thank David Pipitone of Lab Safety Supply Company and the staff members of the U.S. Consumer Product Safety Commission's Chemical Hazards Program for providing the full results of their surveys for this summary.

REFERENCES

1. *Safety in Academic Chemistry Laboratories,* 4th ed. (1990). American Chemical Society Committee on Chemical Safety, Washington, DC.

2. Safety Committee of the New York Section of the American Chemical Society (1981). "Guidelines for a Complete Safety Audit in the Chemistry Laboratory," *J. Chem. Ed.* **58**, A361.

3. Safety Committee of the New York Section of the American Chemical Society (1982). "Results of Safety Inspections of College Laboratory and Chemical Storage Facilities," *J. Chem. Ed.* **59**, A9.

4. D. A. Pipitone and D. D. Hedberg (1982). "Safe Chemical Storage: A Pound of Prevention Is Worth a Ton of Trouble," *J. Chem. Ed.* **59**, A159.

5. R. S. Fausett (Jan. 5, 1982). "Status Report: School Laboratory Chemicals," Memorandum to the United States Consumer Product Safety Commission, Washington, DC.

6. R. L. Schmidt (1977). "Academic Experiences with O.S.H.A.," *J. Chem. Ed.* **54**, A145.

7. *Federal Register* **39**, No. 125, June 27, 1974.

8. N. V. Steere (1967). "Fire-Protected Storage for Records and Chemicals," *Safety in the Chemistry Laboratory,* Vol. 1, Division of Chemical Education of the American Chemical Society, Easton, pp. 44–47.

9. H. L. Jackson, W. B. McCormack, C. S. Rondestvedt, K. C. Smelz, and I. E. Viele (1970). "Control of Peroxidizable Compounds," *J. Chem. Ed.* **47**, A175.

10. D. Pipitone (1981). "Safe Storage of Chemicals," *The Science Teacher* **48** (2).

11. A. J. Shurpik and H. J. Beim (1982). "A Chemist's View of Labeling Hazardous Materials as Required by the U.S. Department of Transportation," *J. Chem. Ed.* **59**, A45.

12. Flinn Scientific Inc., P.O. Box 231, 910 West Wilson Street, Batavia, IL 60510; *Chemical Catalog Reference Manual,* issued annually.

13. M. E. Green and A. Turk (1978). *Safety in Working with Chemicals,* MacMillan, New York.

14. N. V. Steere (1971). *Handbook of Laboratory Safety,* 2d ed., Chemical Rubber Publishing Company, Cleveland, OH.

15. F. W. Mackinson, R. Stricoff, and L. J. Partridge, Jr. (Eds.) (1978). *NIOSH/OSHA Pocket Guide to Chemical Hazards,* U.S. Government Printing Office, Washington, DC.

16. N. I. Sax (1979). *Dangerous Properties of Industrial Materials,* 5th ed., Van Nostrand Reinhold, New York.

17. Laboratory Safety Supply Co., P.O. Box 1368, Janesville, WI 53545; Catalog No. 20-2011 (Pocket Size, Catalog No. 20-2012).

18. Division of Training and Manpower Development (NIOSH) (1980). *Manual of Safety and Health Hazards in the School Science Laboratory,* U.S. Department of Health and Human Services, Cincinnati, OH.

CHAPTER 11

UNIVERSITY OF AKRON
CHEMICAL STORAGE FACILITIES

FRANK L. CHLAD
Laboratory Design Consultant
F.L. Chlad & Associates
Aurora, Ohio

11.1 SAFETY AND FACILITIES

One of the greatest problems facing chemistry departments is that of providing adequate and safe chemical storage facilities. All too often this is an afterthought, rather than being an integral part of the basic planning and design of a new building.

The central storage facility for chemicals must be carefully thought out and well planned, and there can be no doubt that it will be expensive if done properly. Among the many critical areas that need to be considered are the following:

1. An adequate space allocation, including future growth potential
2. An efficient ventilation system
3. Heat and smoke detectors
4. An automatic fire suppression system
5. Explosion-proof lighting (and any motors)
6. Static-free light switches and electrical outlets
7. Proper temperature and humidity controls
8. Pressure-releasing plastic blow-out panels
9. Fire alarm and annunciator system

One must not view chemical storage facilities as some type of appendage, separate and distinct from the chemistry building itself. Rather, they should be an extension of the safe design concepts that must be planned into the total building. The ultimate goal is to plan a totally safe building, with equal consideration given to the chemical storage facilities during the original planning and design.

For purposes of illustration, this author will utilize the facilities at the University of Akron (Akron, OH) as an example of the total safety design concept.

Knight Chemical Laboratory, which houses the Department of Chemistry at the University of Akron, has received worldwide acclaim as being one of the safest academic chemistry buildings ever constructed. Now in its tenth year of operation, it remains a prototype for illustrating what can be accomplished when safety is the dominant theme in designing chemical facilities. Well over 100 inquiries have been received from all over the world, and thus far 38 institutions have sent representatives to view this structure (Figure 11.1).

11.2   LABORATORY FACILITIES

One of the more important of the many innovative safety features incorporated into the laboratories was the installation of 166 induced-air fume hoods, which makes it possible for *all* experiments to be conducted in hoods.

FIGURE 11.1   Knight Chemical Laboratory at the University of Akron in Ohio. External fume hood exhaust chases are shown.

To conserve energy, the hoods take only 25% of the air from inside the rooms. The remaining 75% is tempered outside air drawn into the building at each floor level. The facility is odor-free because chemicals fumes are drawn out of the building at approximately three times the rate of exhaust obtained with a standard-type ventilation system.

11.2.1   Undergraduate Laboratories

Generally, the undergraduate organic laboratory poses the largest potential danger as far as accidents, fire, and noxious fumes are concerned. This is due to both the nature of the organic chemicals utilized and the use of large quantities of solvents. An in-depth study of accidents ascertained that the largest number of accidents occurred in the undergraduate organic laboratory. These incidents ran the full spectrum from minor fires caused by exposing solvent vapors to bunsen burners, to chemical burns caused by carrying the materials from a laboratory bench to the hood, and bumping into someone. Our new design and procedures have improved the situation tremendously.

The organic laboratory is divided into two areas served by a common instrument room. Each of the two laboratories contains sixteen 8-foot auxiliary air hoods at which the students do *all* of their work. There are two student stations per hood, giving a total of 64 working stations at any given time. Because all work is done in the hoods, rather than at benches and hoods, the previous congestion and safety problems associated with this laboratory have been greatly minimized (Figure 11.2).

Because of the large number of hoods in the two undergraduate organic laboratories, a tremendous amount of outside air is brought in to maintain the 75:25 ratio that is exhausted. As an example, rooms of this size without the hoods would normally require 4–6 air changes per hour, whereas under the system designed for us, the laboratories receive 109 air changes per hour, or one every 33 seconds.

Several other factors are noteworthy. The organic laboratory is now flameless since bunsen burners have been replaced with heating mantles, steam baths, and hot plates. We have converted to 19/22 standard taper semimicro glassware kits for all students. This has reduced the quantities of materials used in experiments, resulting in an economic and safer operation.

Compare, if you will, the obvious safety features described above with the situation that prevailed in the "classic" organic laboratory. The undergraduate laboratory, we all too vividly recall, served 24–36 students and forced the students to share four to six ill-working hoods that were normally located against one wall. For each person working in the hood there were three or four waiting in line, all carrying their materials and chemicals from the bench to the hood and back again. Many did not bother waiting for the hood to be free and did their work on open benches. Solvent fires caused by vapors reaching a lit bunsen burner were all too common. Designing safety into the planning of these laboratories has allowed us to make tremendous progress.

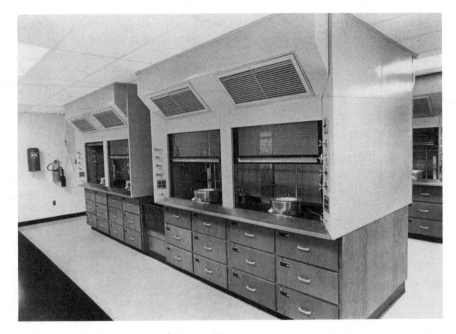

FIGURE 11.2   A standard undergraduate organic chemistry laboratory. All experiments are conducted in fume hoods.

The concept that all chemical work must be performed in a controlled environment is also carried out at the freshman level. There are eight introductory chemistry laboratories arranged in four blocks of two, with each pair of two separated by sliding pocket doors. A laboratory has 24 student stations, each one with a bench-top T-type hood. The hood has an adjustable baffle to allow for the removal of vapors heavier or lighter than air. Each laboratory has, in addition, an induced-air fume hood from which chemicals are dispensed during each laboratory period. The 24 T hoods are exhausted as a group through a common duct, whereas the storage hood is exhausted separately (Figure 11.3).

11.2.2 Graduate Laboratories

Much consideration was given to the design of the graduate research laboratories. Each laboratory is 24 × 30 feet and is designed for occupancy by four students. It was felt that installation of a series of two-student laboratories would result in a needless duplication of equipment, whereas placement of more than four people in a single room would lead to overcrowding and a lack of identifiable space. Our experience with hood usage in our old building revealed that most experiments were carried out on bench tops because the

FIGURE 11.3 Introductory chemistry laboratory illustrating the individual T-type hoods at each station. In case of emergency, each laboratory can be quickly evacuated.

hoods were used primarily for chemical storage. This practice had the further disadvantage that it required the hoods to be left on continuously. Our solution to this in the new facility was to place five hoods in each four-student laboratory. Four of the hoods are used to conduct experiments, whereas the fifth hood is used solely for storage. The four "working" hoods are turned on only when in actual use, and the storage hood is left on continuously. The storage hood does not have any utilities piped to it and is fitted with shelves. The net result is that the working hoods are not cluttered with chemicals and are turned on only when a reaction is running. Thus, conditions are safer and energy is conserved as well.

The location of the graduate research student desks within the laboratory is vitally important. The ideal situation is to have desks located against interior walls to allow quick egress to the hallway. The hoods are located against exterior walls so that ducting into chases on the outside of the building may occur. These measures allow a student to exit from the laboratory quickly in an emergency situation and also eliminates the necessity of having other students or visitors walk through a potentially dangerous area to reach a graduate research student for conferences. The location of the hoods against the outer wall also eliminates the turbulence that would be caused by opening and closing doors and when persons walk continually past the hoods.

Each four-student graduate research laboratory has a "safety island" consisting of an approved solvent storage cabinet, fire extinguisher, water sprayer, bucket of sand, eyewash station, and spill kit (Figure 11.4). Another key feature in the research laboratories is that the utilities in each individual laboratory are able to be shut off from a control panel located in the corridor directly outside each room, so that in case of fire or other accident any laboratory can be quickly isolated from the others (Figure 11.5).

11.3 FUNCTIONAL CHEMICAL STORAGE

Because of the diverse nature of chemicals and their use, the safe storage of chemicals has become a matter of paramount importance. Ideal hazardous chemical storage would entail the complete isolation of each hazard category and even isolation of some materials within a class. However, from a practical standpoint, such isolation is not often economically feasible, and it is thus necessary to group items so that whatever space is available is used in the safest possible manner.

Designing safety into chemical storage facilities involves a great deal more

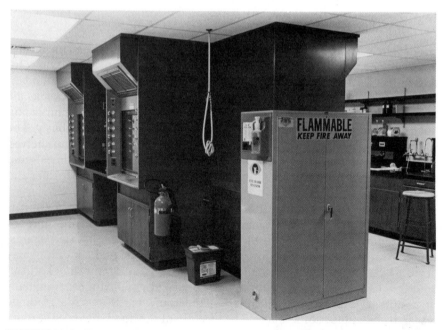

FIGURE 11.4 Four-student research laboratory featuring back-to-back work hoods, a storage hood, and a safety island.

than sketching out a proposed floor plan and specifying certain safety "hardware." It requires a careful examination of the total concept of chemical safety from a logistical standpoint, with a complete overview of what happens to a chemical from the time it is received at the loading dock, until ultimate consumption, or disposal of the chemical as a waste.

11.3.1 Chemical Storage and Undergraduate Laboratories

In all too many instances the following examples are common practice. A chemical ordered for a particular undergraduate teaching laboratory is received and is placed in that teaching laboratory until it is used. Each undergraduate laboratory thus becomes a "ministockroom" with little or no control procedures. Particularly in the case of the undergraduate organic laboratories, the toxic or noxious chemicals are stored in fume hoods. The fume hoods become storage areas rather than working areas. Normally, there is not a sufficient number of hoods in which to work.

Another situation occurs frequently when experiments or laboratory manuals are changed in the curriculum. Many chemicals that are no longer needed as a result of the change are still stored in the laboratory with the

FIGURE 11.5 An access closet located outside each laboratory contains utility shutoff valves for emergencies.

thought that undoubtedly usage will occur "sometime" in the future. This causes untold problems with storage space, the reliability of the chemical, and ultimately, a costly disposal situation.

The Department of Chemistry at the University of Akron has adopted a policy that has worked exceedingly well. Absolutely no chemicals are stored in undergraduate laboratories. Instead, the department has several dispensing stockrooms that service the various teaching laboratories by means of a cart system. The system works in the following manner. The dispensing stockroom storekeeper requisitions those chemicals that will be needed for the following week's laboratories from the main chemical storage area. The storekeeper then prepares the needed unknowns, solutions, and materials that will be required. The items are placed on laboratory carts, each identified by the course and laboratory room number. The graduate teaching assistants obtain these chemicals at the dispensing stockroom five minutes prior to the start of the actual laboratory period.

These materials are wheeled into the individual laboratories and are then issued to the students. At the end of the laboratory period any unused materials or waste is loaded onto the cart and returned to the dispensing stockroom. Several objectives are accomplished by utilizing this system: (a) there is a high degree of control exercised over the use of the chemicals; (b) laboratory maintenance is excellent since chemicals are absent in the laboratories when not in actual use; and (c) the administration is able to analyze accurately the cost of operation for each course. The end result is an uncluttered laboratory with no storage problems and, most importantly, a safe operation.

Our philosophy is based on the premise that if one has an excellent chemical storage facility, with all the many safety features mentioned at the beginning of the chapter, one should store the chemicals in that area for as long a period of time as possible prior to their actual use.

11.3.2   Chemical Storage and Graduate Laboratories

The problem of chemical storage in research laboratories is also a major concern. In many respects this is even more complex because one needs to deal with a host of faculty conceptions, each presenting a varying degree of safety consciousness. In addition, some faculty with funded research often feel that they are independent of any department system of safety or inventory control. The department must, however, have a firm policy of disallowing private caches of chemicals in the research areas and should limit use to only those chemicals that are currently being utilized. Everyone should follow one set of guidelines, with equal enforcement.

All too frequently, a research group will sign out a 1-lb bottle of a substance, use 50 g in a reaction that does not work, and then store the remainder on a shelf in their laboratory for years. One can easily envision the large quantity of chemicals that can be accumulated over a short period of time. Laboratories should be work areas and not storage areas.

Our solution to this situation was the establishment of a chemical "morgue" that operates in the following manner. Once a chemical is determined to be in excess, it is brought to the main chemical storage facility where it is properly catalogued and placed in the chemical morgue. This morgue area is distinctly separate from the regular chemical stock and is kept on a different inventory system. Any faculty member or graduate research student who has a need may sign for this chemical at no charge. In this manner we are able to recycle all excess chemicals and keep supply costs to research grants and the department budget at a minimum. Storage conditions are improved from both a safety and space standpoint, and ultimately waste chemical disposal is kept to the lowest possible level.

The use of a morgue system does not have to be limited to only academic institutions. The author has had occasion to visit many industrial research laboratories. A good many of these were cluttered and underhooded and had unsafe quantities of chemicals stored in their work areas. These conditions represent very high potential for disaster.

What is being discussed here is the classic trade-off between convenience and safety: the convenience of having that "ministockroom" in the research laboratory with an unsafe condition, or the willingness to make a few extra trips per week to the central storage facility, with a safe situation. The amounts of toxic, flammable, unstable, or highly reactive materials that should be allowed to be stored in individual laboratories must be governed. Unrestricted quantities of these materials can bring about untold safety problems and hazards. Ground rules should be established and then conscientiously enforced. An excellent reference for determining the maximum quantities of flammable and combustible liquids in laboratory units outside of approved flammable liquid storage rooms is NFPA booklet No. 45-198, *Fire Protection for Laboratories Using Chemicals,* prepared by the National Fire Protection Association.

11.4   CHEMICAL STORES AND WASTE DISPOSAL

With the advent of the Resource Conservation and Recovery Act, tighter control of chemical inventory and the subsequent utilization of this morgue system has become a matter of extreme importance. When our chemistry department moved from the old building to our new facility, the magnitude of the problem of poor inventory control of chemicals became very evident. Many of the laboratories, both teaching and research, had a 15–20-year accumulation of chemicals. A determination had to be made as to whether these chemicals could still be utilized or whether disposal was needed. Approximately 25% of the chemicals that had been stored were discarded. In this group, chemical containers had no labels or labels that could no longer be read, container caps had corroded completely, and chemicals had become contaminated.

The identification process and ultimate disposal was a nightmare. The cost in work-hours and dollars, in addition to the highly hazardous conditions that were uncovered, were definitely instrumental in the formulation of our new policies and procedures.

Another situation resulting from the Resource Conservation and Recovery Act caused an additional problem situation. Our department was deluged with the offer of "free" chemicals. A host of companies and high schools contacted our department to offer "gifts" of chemicals claimed to be in excess to their operation. In reality, they were trying to unload old, and in many cases contaminated, chemicals that they realized needed to be discarded. By offering these chemicals as gifts, these people not only hoped to avoid high disposal costs but were also seeking a tax deduction for a charitable contribution. We set a policy for accepting donations of chemicals that includes a stipulation that we be allowed to select only those that we can utilize, that the chemical be in the original container, that a proper label be affixed, and that we are certain that the chemical is pure.

With the host of edicts from regulatory agencies, namely the EPA, the OSHA, and local and state governments, chemistry departments must have established policies and procedures for the handling of their chemical waste. At the University of Akron waste solvents are handled in the following manner. Each research laboratory is issued a 5-gallon approved safety can that is kept in the solvent safety cabinet. A tag is wired to the handle of the safety can. Each time a person pours waste solvent into the safety can, the tag is updated with the person's name, the type of solvent, the quantity, and the date. When the 5-gallon safety can is full, it is delivered to the chemical stores area, whereupon the researcher receives an empty container and a new tag. The full container is transferred into an approved 55-gallon drum (utilizing a safety funnel with flame arrester and bonding wires). The information from the tag for the 5-gallon container is transferred to the 55-gallon drum manifest. In the case of very expensive solvents, some of the research laboratories choose to redistill their waste solvent. An 80% recovery rate has been reported and the redistilled products are often purer than the original start-up material.

Chemicals other than waste solvents that require disposal are brought to the chemical stores area. The chemical names and quantities are recorded on a manifest. Storage takes place in a special holding room within the chemical stores complex. According to law, the department cannot store waste chemicals for more than 90 days without a special permit from the EPA as a storage facility. Therefore, a minimum of four shipments per year are made to an EPA-approved disposal agent. Waste solvents are shipped in approved 55-gallon drums (bung type). Other chemicals must be separated into appropriate categories (e.g., organics, inorganics, active metals) and packed in layers of absorbent packing material in separate open-head 55-gallon drums. Each drum requires the appropriate labels as mandated by the EPA and the Department of Transportation. A shipping manifest that indicates the identity and quantity of each chemical accompanies each drum.

The following are some helpful recommendations regarding chemical waste:

1. Exercise a grat deal of caution with regard to the type of experiments chosen for teaching laboratories. Know in advance what kind of waste will be generated. Careful thought in the selection of a laboratory experiment may prevent costly and time-consuming disposal problems later on.
2. Establish a written policy regarding chemical waste disposal and insist that it is followed. Accurate records are mandatory.
3. Be aware of the fact that there are ways to dispose of much of the chemical waste generated without the necessity of shipping it to an approved disposal agent. See "Hazardous Chemical Waste and the Impact of R.C.R.A.," *J. Chem. Ed.*, 331–333 (April 1982).

The author's purpose in describing the above details lies in the highly complex nature of hazardous waste disposal since the advent of the EPA regu-

FIGURE 11.6   The main chemical storage facility. The building is earth-banked on three sides, and the roof contains pressure-releasing plastic blow-out panels.

lations published in late 1980. The impact on the economic and logistical operation of all chemistry departments has been great. In designing new chemical storage facilities, or in the renovation of existing ones, it is crucial that these factors be kept in mind. Proper storage and handling facilities for hazardous waste must be well thought out and included in the overall storage scheme.

In the Knight Chemistry Laboratory at the University of Akron, the chemical stores are contained in a one-story wing separately connected to the main chemistry building. Immediate access to a freight elevator and loading dock is provided. The wing is landscaped on the three exposed sides with an earthen bank that extends to within six feet of the roof line. The roof itself is equipped with ventilation fans and pressure-releasing plastic blow-out panels (Figure 11.6).

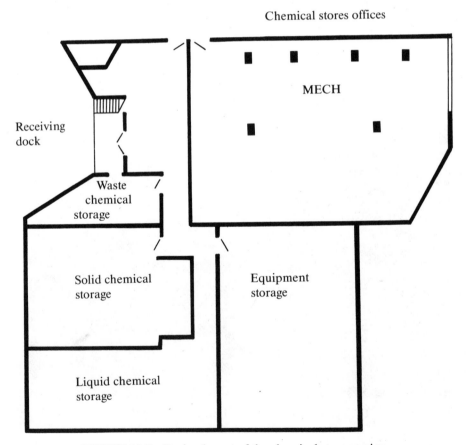

FIGURE 11.7 Design layout of the chemical storage wing.

FIGURE 11.8 An annunciator system that uses a lighted panel to pinpoint the location of fire while sounding an audible alarm.

FIGURE 11.9 Static-free switches and electrical receptacles reduce fire danger.

Inside the building, space is divided into a receiving room, a waste solvent holding area, and three separate storage rooms for dry chemicals, liquid chemicals, and chemical equipment, including glassware (Figure 11.7). Within the chemical storage rooms, materials are grouped generally alphabetically but overriding consideration is given to compatibility.

If a fire should break out in any of the storage areas, the smoke and heat detectors will trigger an alarm, and 30 seconds later the affected area will be blanketed with carbon dioxide. At the same time, an annunciator system located in the chemicals stores office will sound an alarm, and, by means of a lighted panel, show the exact location of the fire (Figure 11.8).

To reduce still further the fire danger, all phones, switches, receptacles, and light fixtures in the chemical stores area are static-free (Figure 11.9).

11.5   CONCLUSION

All too often, the provision of adequate chemical storage space is given little consideration by university administrators when viewed as a "nonproductive" aspect of the operation. However, the lack of sufficient storage space can create hazards due to overcrowding, storage of incompatible chemicals together, and poor housekeeping. Adequate, properly designed, maintained, and well-ventilated chemical storage facilities must be provided to ensure personnel safety and property protection.

Those involved in chemical safety have a tremendous responsibility to advocate safe chemical storage facilities. University administrators and top management in industry must be convinced that chemical storerooms, although costly when done properly, are critical to the safe operation of a modern chemistry facility. To inculcate decision makers that storerooms are not to be added later as an afterthought or construed as a dead space in which shelves are erected is only part of the task.

There is a saying among architects that "design follows purpose." This is especially true when designing safe chemical storage facilities. No amount of brick and mortar, regardless of how arranged, will make an operation completely safe unless a total program of safety is present. Safe storage facilities are merely one ingredient in the total picture.

Although the safety designed into a building is vitally important, in the final analysis, people will be the determining factor as to the overall safety of the operation. People have within themselves the potential to either prevent or cause an accident. Chemical safety involves much more than sophisticated exhaust systems and careful design and layout of laboratories and storage areas, as well as the many facets of structural detail that go into designing a building. People must be taught how to be safe.

A serious and dedicated commitment to safety, with expenditure of time, energy, and money can provide a safe environment in which to work, teach, and learn.

CHAPTER 12

SAFETY FIRST IN THE ELIMINATION OF HAZARDOUS CHEMICALS FROM SCHOOL SCIENCE SETTINGS

JACK GERLOVICH
Assistant Professor
Drake University
Des Moines, Iowa

12.1 INTRODUCTION

The intent of this chapter is to assist school personnel in making science environments safer places in which to teach and learn. This is best accomplished by helping identify and eliminate hazards and by providing information relative to a proven and cost-effective plan for disposing of unwanted chemicals.

It was a bright winter day in the upper Midwest. Tom and John were anxious to get to science class—their favorite ninth grade subject. Room 142 always reminded them of some disorganized scientist's laboratory, as apparatus, open containers of all types of chemicals, and half-finished experiments lay everywhere.

Tom and John had discussed the experiment they were going to perform. They knew that picric acid would fit their extracurricular scientific needs very nicely.

They met Mr. Thomas, their science teacher, at the doorway of the class. He said that he had to attend to some office work and would return in 20 minutes, however, they could work in the laboratory during that time. Be careful, he added, and then disappeared down the hall, papers in hand.

John and Tom proceeded to the rear of the classroom/laboratory to the open chemical storage shelf area. The picric acid was on the top shelf. Tom secured the rickety ladder used for placing and retrieving such chemicals from the shelves. They had performed this task many times in their two-year junior high science education process. Tom climbed the ladder and lifted the two-pound container of crystalline picric acid from the shelf. They placed the container on the counter and proceeded to open it. As John unscrewed the container lid, an explosion ensued, injuring both boys and seriously damaging the laboratory.

Could the accident have been prevented? Could such chemicals have been safely and cost-effectively removed from the school to prevent such incidents?

12.2   THE PROBLEM OF UNWANTED CHEMICALS

The great majority of schools in this nation are stockpiling unwanted and often hazardous laboratory chemicals due to the prohibitive costs and government regulations required to legally dispose of them.

In 1986 and 1987, many states implemented Hazardous Chemicals Risk Right to Know legislation designed to make places of employment safer. Partly in response to the 1984 chemical accident in Bhopal, India, and the ensuing public pressure, Congress passed the Superfund Amendment and Reauthorization Act (SARA) in 1986. SARA Title III, also known as the Emergency Planning and Community Right-to-Know Act require that employers identify all hazardous chemicals, label them properly, and apprise employees, the public, and emergency services agencies of their existence, associated hazards, and appropriate emergency procedures. Specifically, local hazard communication programs must provide:

1. Transmission of information to employees concerning hazardous chemical products to which they may be exposed in their workplace.
2. A clear mechanism for providing information to local emergency response organizations relative to hazardous chemical products stored at the school.

These two steps are best accomplished by developing a complete inventory of chemical products stored and used in the science program, and then determining which are hazardous. In the Iowa Legislation [530-130(455D)], for instance, a chemical is hazardous, in any quantity, if the National Fire Protection Association (NFPA) chemical hazard rating is greater than 3 for health, flammability, or reactivity, or if 25 pounds, or 250 gallons aggregate of chemicals have NFPA ratings of 2 or greater for flammability, reactivity, or health hazard.

All chemical containers must be properly labeled with the following information:

Chemical name
Appropriate warnings
Name and address of manufacturer or distributor

All buildings smaller than 5000 square feet containing hazardous chemicals will have NFPA placards on the outside doors. For buildings of greater than 5000 square feet, placards will be placed where chemicals are stored. Hazard placards are NFPA chemical hazard diamonds.

Material Safety Data Sheets (MSDSs) must be available as the official reference to use in ascertaining chemical facts, appropriate first aid, and emergency reactions. All MSDSs must be placed in convenient places where they would be immediately accessible to the user (teacher, custodian, etc.). Provisions should be made by the district purchasing department to require that all chemicals purchased be accompanied by MSDSs.

Training must be provided to all employees relative to chemical hazards. Specifically, instructions should include the use and location of MSDSs; recognizing chemical hazards; and interpreting labels, signs, color codings, and appropriate emergency reactions. When new chemicals are introduced, similar training should be provided relative to them. Permanent individual records of such training should be maintained, including the trainee's name, signature, date of training, and trainer.

Lastly, a written copy of the Hazard Communication Program plan must be provided to employees, designated government agencies, and emergency response teams (fire departments, rescue squads, hospitals, etc.). It is best to have the plan reviewed and co-signed by the local fire department or regional fire marshall prior to submission.

In many schools chemicals stored are unidentifiable due to age or container abuse. In most instances, school personnel have decided it would be best to purge the chemicals rather than attempt right-to-know compliance for them. However, at about this same time, the EPA tightened all RCRA regulations concerning the identification and disposal of hazardous waste for small-quantity generators, such as schools. In some instances, this situation was complicated even further when insurance carriers would not cover schools attempting to remove hazardous chemicals from their premises. The announcement below, for example, was sent on December 30, 1988, from the Iowa Department of Education and the state's largest school insurance carrier to all intermediate and local education agencies (schools) in Iowa.

"Don't Let Hazardous Chemicals Blow You Away"

Hazardous chemicals are not just in our school laboratories. They are in our janitorial closets, our garages, and our boiler rooms. Although these products are useful and necessary, they pose a variety of problems to schools. If not handled and stored properly, they threaten the health and safety of employees and students, school neighborhoods, and the environment.

The Iowa Right-to-Know Rules require employers, including schools, to develop and implement a written hazard communication program for the workplace. This program includes the use of labels, notices and other forms of warning as well as inventory control and recordkeeping. The purpose is to make the employee and the employer more knowledgeable about the products with which they work, thus making the workplace safer. Civil penalties may reach $10,000 per violation for willful violations of these provisions; criminal penalties are also applicable in some cases.

Aging and out of date chemicals pose special problems because their characteristics change. Although employees may know how to properly handle fresh chemical products, they cannot judge the volatility, reactivity, or the overall threat posed by old products. The risk of fire and explosion become greater.

At this point, these chemicals are waste products and most should be considered hazardous waste. Hazardous waste must be disposed of in accordance with federal and state environmental regulations. Failure to manage wastes properly may result in civil penalties of $25,000 per day per violation and/or criminal penalties. Using a reputable and experienced contractor to remove and properly dispose of the waste helps reduce the risk to school personnel and the environment as well as reducing liability exposure. Any contract for removal and disposal should be reviewed by legal counsel for the district before signature. The contractor should be prepared to provide a Certificate of Liability Insurance and Workers' Compensation Insurance, as well as assurances that removal and disposal will proceed pursuant to applicable federal and state law.

Recently, the insurance industry has reminded all Iowa school boards that school liability insurance does not cover pollution events. Because removal of hazardous chemicals from our schools may create a pollution potential, school personnel find themselves between a rock and a hard spot.

Nonetheless, maintaining old and unused hazardous chemicals in our schools appears to pose an even greater threat.

Given the hazardous nature of these chemicals, it is our feeling that schools should proceed to remove old and unused hazardous chemicals from schools. Special care should be taken to use proper personnel to conduct the work. The chemicals must be disposed of pursuant to state and federal regulations and contractors engaged in this work must provide documentation of authorized disposal practices.

The question thus became not *whether* to dispose of chemicals, but rather *how* to best manage and dispose of those identified as hazardous in the safest, most cost-effective manner.

12.3   THE IOWA MODEL

The Iowa plan was based extensively on safety, and it is assumed that anyone attempting to use the plan will also emphasize the need to be safe when preparing for and handling such chemicals. In addition, the plan was cost-effective due primarily to the involvement of all schools and the ability to pick up chemicals at regional sites.

12.3.1   Forming an Advisory Committee

Due to their educational responsibility, regular interaction with educational agencies, and other state departments, the state department of education should serve as the coordinator of the overall project for the statewide safety and chemical disposal operation. For parallel operations at the county or regional level, a similar umbrella educational agency should coordinate the effort. The following agencies should form the nucleus of the advisory committee:

State department of education
State department of natural resources
State department of transportation
State fire marshall's office
Intermediate educational agencies (if applicable)
University chemist
Classroom chemistry teacher
School administrator
School board association
Major school insurance carrier

12.3.2 Developing an Action Plan

The advisory committee should anticipate approximately 12 months to complete the safety assessments, corrections, and chemical disposal operation:

Time Required	Actions Completed
3 Months	Develop and refine total plan
1 Month	Inservice of participants
2 Months	Participants perform safety, assessment, and chemical inventory
6 Months	Chemical redistribution, pickup, and disposal

The advisory committee should complete preliminary plans and secure tentative approval for them from necessary regulatory state and federal agencies. Regular consultation should occur with the state departments of transportation and natural resources and the fire marshall's office to ensure compliance with transportation, disposal, and chemical destruction guidelines. The fire marshall's office, for instance, remotely detonated many of Iowa's explosive chemicals. This helps reduce the chemical pool destined for disposal.

Pay careful attention to the selection of the chemical transport and disposal company selected for these tasks. Check with your state department of natural resources concerning companies that have experience in doing this type of work. Be careful to eliminate those who have committed environmental violations. Most important, the advisory committee should ask each company bidding on this job, to put its plan in writing (including costs). After review of the written plans, submitted by the disposal companies, the strongest contenders should be invited to present, in person, to the advisory committee. *It is critical that the company selected have past experience in such operations with schools! It is equally critical that the company have several disposal options* (such as landfill sites, incineration sites, or redistribution options).

Bear in mind that landfilling of chemicals leaves liability for the chemicals with the generator (school) in perpetuity. Incineration, by contrast, allows such liability to "go up in smoke." Incineration costs, however, are generally 50–70% higher than is landfilling. In addition, since only approximately 60% of typical school chemicals have sufficient BTU value to be incinerated, landfilling and redistribution will need to be options. Check with your selected transport and disposal company for details.

Mandatory inservice programs for personnel from all participating educational agencies should be developed as part of the overall plan. The inservice must include a safety prerequisite as well as the mechanics of the pickup process.

Following necessary modifications, the plan should be forwarded to the respective regional EPA office (see Table 12.1 for suggestions and approval).

12.3.3 Planning the Inservice Program

12.3.3.1 Inservice/Chemical Pickup Sites
The inservice should be one-half to one full day in length. Teams of chemistry teachers and principals from each district participating in the cleanup operation should be required to attend the event.

The inservice site should be at a location relatively central to all participants in the designated regions of the state. Intermediate educational agency offices, community colleges, and college and university campuses are convenient and effective locations. The inservice sites should be familiar to the participants, have an educational affiliation, and be able to accommodate the size of the group anticipated for that region. In addition, the inservice site should be the future pickup site for the chemicals. The facility will need to have a covered space that will accommodate semi-trucks. This will enable dry transfer of the chemicals from local school vehicles to transport "lab packs" (55-gallon barrels with sufficient inert packing material to absorb all chemicals packed within) in the event of precipitation. The area should be conveniently cordoned off from traffic, have telephone access to the outside, and accommodate cars driving up close to the repacking site. There will also need to be a large open area where the floors can be covered with inert material (dried clay, vermiculite, etc.) and where the chemical disposal/transport company can repackage the chemicals into "lab-packs" for transport. Last, the area will need to accommodate a large, commercial dumpster into which old packaging material can be disposed.

12.3.3.2 Safety
Speakers should be identified who are knowledgeable of science safety and who have been involved in school chemical pickup operations in the past. It is imperative that safety be the focus of the entire effort!

Tort law should be discussed, including negligence and liability, goggle laws, insurance coverage, and right-to-know legislation unique to the state. In

TABLE 12.1   Environmental Protection Agency Offices

EPA/RCRA Hotline, 1-800-424-9346 (in Washington, DC, 382-3000)

EPA Region 1
JFK Federal Bldg.
Boston, MA 02203
617-223-7210
States served:
CT, MA, ME, NH, RI, VT

EPA Region 2
26 Federal Plaza
New York, NY 10278
212-264-2528
States served:
NJ, NY, PR, Virgin Islands

EPA Region 3
841 Chestnut St.
Philadelphia, PA 19107
215-597-9800
States served:
DE, MD, PA, VA, WV, DC

EPA Region 4
345 Courtland St.
Atlanta, GA 30365
401-881-4727
States served:
AL, FL, GA, KY, MS, NC, SC, TN

EPA Region 5
230 S. Dearborn St.
Chicago, IL 60604
312-353-2000
States served:
IL, IN, MI, MN, OH, WI

EPA Region 6
1201 Elm St.
Dallas, TX 75270
214-767-2600
States served:
AK, LA, NM, OK, TX

EPA Region 7
726 Minnesota Ave.
Kansas City, KS 66101
913-236-2800
States served:
IA, KS, MO, NE

EPA Region 8
One Denver Pl.
999 18th St., Suite 1300
Denver, CO 80202
States served:
CO, MT, ND, SD, UT, WY

EPA Region 9
215 Fremont St.
San Francisco, CA 94105
415-974-8071
States served:
AZ, CA, HI, NV, American Samoa,
Guam Trusts of the Pacific

EPA Region 10
1200 Sixth Ave.
Seattle, WA 98101
206-442-5810
States served:
AK, ID, OR, WA

addition, fire codes, OSHA regulations, and science standards of professional societies should be presented. These items, in conjunction with a review of appropriate accident case study reviews, will help participants gain a broader perspective of their professional, moral, and legal responsibilities toward student and peer safety.

A software program was used to communicate legal concerns and to provide a standardized, automated process for conducting a safety assessment of school facilities, equipment, and practices. Safety checklists were included that enabled school personnel to assess the safety of classrooms, laboratories, teacher preparation rooms, storerooms, and field environments. Forms were also included that enabled the school personnel to initiate corrections of identified safety hazards with the administration. Accident/incident reports and student safety contracts were also provided. This software program was invaluable in ensuring that necessary safety equipment, such as fire extinguishers, fire blankets, eyewashes, cover goggles, drench showers, exhaust hoods, and intercom/telephones, were available in the proper quantities and sites; that equipment was of the proper type and size; and that such items were functioning properly. In addition it ensured that the environment was safe from "foreseeable" hazards *before* the chemical redistribution and disposal operation was initiated. *Participating educational agencies were strongly advised to perform the total safety assessment before initiating any work with chemicals.*

12.3.3.3 *Chemical Management*   The software program also provided a standardized, automated inventorying, labeling, storage, and hazard assessment system for typical school science chemicals. The program provided a database of 420 chemicals commonly used in secondary school and community college science programs. Included, for each chemical rated, was information concerning NFPA Hazardous Chemical Codes, DOT chemical hazard categories, chemical synonyms, chemical family storage information, chemical storage plans, and chemical fact sheets. A total of 700 chemicals could be contained on the disk if all fields were filled.

The safety and prepickup chemical management components were essential to the success of this operation. No compromises should be tolerated relative to these steps.

The Iowa State Board of Education, in cooperation with a major school insurance company, supplied each participating school with a set of the science safety software discussed above.

12.3.3.4 *Transport/Disposal Company Input*   Through a modified bid process, SET Environmental, Inc. (Wheeling, Il) was selected from among 10 companies to do the chemical transport/disposal operation. SET had the experience and qualified personnel to do the job. It was also the only company to guarantee prices in writing. Instructions were also provided by SET, Inc., personnel in completing EPA temporary disposal permits. This greatly reduced the fear of participants in properly completing the necessary federal paperwork. The form designed by SET was streamlined and logical to use.

Instructions were also provided by SET concerning the packaging of chemicals identified for transport and disposal to the regional pickup site. This step was invaluable in meeting the concern of school personnel for transporting potentially hazardous chemicals in school vehicles.

Costs involved in the pickup and disposal operation was also provided. There was *no* pressure exerted on schools to participate in the process. Obviously, however, the more schools that participated, the greater the amortization of such costs could be.

Lastly, the timeliness for completing each of the above steps was provided. In many instances, school personnel felt that the process should have proceeded faster. It was felt, however, that haste would only compromise safety.

Following the inservice program, participants were ready to return to their individual schools to perform safety assessments and the inventorying of chemicals.

The automated, standardized process greatly facilitated the chemical clean-up. All environments could be assessed for hazards, and corrections could be made prior to chemical movement. Inventories could be standardized, thus enabling comparison across districts and ease of management by state agencies and the transport/disposal company. The complex, time-consuming process of chemical inventorying and management was greatly facilitated. Questions across districts were easily addressed, since all participants were using the same system. Inservice training effectiveness was also greatly enhanced.

12.3.4 Conducting the Chemical Redistribution and Disposal

Following performance of the safety assessment, school personnel inventoried their current chemical storerooms, and identified chemicals they wished to dispose of. In most cases chemicals slated for disposal were those that were rated as hazardous, those that schools had in excess quantity, or those with incomplete or missing labels.

Inventory sheets (Figure 12.1) were developed by SET, sent to participating school personnel for completion, and then returned to SET. The forms proved to be simple yet very effective in meeting federal requirements, while at the same time alleviating fears of school personnel.

SET analyzed the inventories and then provided each school with packaging instructions and the cost for the redistribution and/or disposal operation. Schools were free, at that point, to withdraw from the program if they desired.

It is important to note that chemicals possess a wide variety of properties and some of them are so uniquely hazardous that they cannot be handled with the general pool. Among those requiring special handling were the following:

Chemical	Handling Requirements
Organic solvents	Require incineration
Water reactives, air-sensitive materials	Require special handling and customized treatment
Heavy metal, salts, cyanides, acids, bases	Require chemical treatment
Unknowns of trade name items	Must be fully identified prior to disposal site acceptance ($60/analysis)
Explosives or peroxide formers	May require remote detonation (most of these were detonated by fire marshalls)

AEA # _____

School Name _____

Initials _____

SET Will Provide This Information:

School Initials and Container Number	Chemical Name	Container Type	Container Size	Amount in Container	Liquid or Solid	Compatibility Code	Drum #	Special Handling Required
PC1	Hydrochloric acid	P	1 qt.	3/4 qt.	L			

SAMPLE

FIGURE 12.1   Sample chemical inventory sheet.

Items requiring extensive special handling (radioactive materials, unknowns, air/water reactives) were addressed separately and were not included in the pricing structure above. These Phase II customized chemical pickups were conducted on site at the respective schools. The cost for customized pickup was lower than if schools conducted the process individually, since the prices were amortized across all Phase II participating schools. Dates were coordinated between SET, Inc. and the schools directly.

Total actual pickup, redistribution, and/or disposal costs were based on the following services, SET agreed to:

Review and code the waste chemical inventories, including EPA and DOT hazard class information.

Organize and consolidate similar and compatible chemicals from all participating schools.

Secure school EPA and DOT chemical transport permits.

Prepare "master drum inventory forms."

Distribute instructions to teachers for safe handling, packaging, and legal shipment of waste chemicals from schools to intermediate pickup sites.

Provide technical specialists, familiar with chemical hazards, spill cleanup procedures, and disposal site requirements to repackage chemicals into approved "lab packs" (55-gallon drums of chemicals and inert materials) for transport and disposal.

Pickup, transport, and dispose of chemicals at licensed site.

On a date designated for their area, schools transported chemicals, packed according to SET's compatibility guidelines from their buildings to the designated area pickup sites. These sites were carefully selected to ensure that driving distances did not exceed approximately 50 miles. Pickup dates were carefully coordinated between SET, the state advisory committee, and local schools.

After the chemicals were accepted at the treatment/disposal site, the following operations were performed:

Manifests were reviewed and checked for completeness and accuracy.

A portion of the drums were subjected to quality control check (comparing drum contents to drum inventory).

All drums were segregated and moved to the respective storage location in the facility.

All drums were logged into a computerized tracking system according to their exact location in the facility.

Permanent records were developed that included detailed tracking of waste processing, treatment, and final disposition of all drums delivered.

Specific treatment, processing, and disposal conducted according to the following generalities:

Item	Treatment
Organic solvents, oils, and other liquids for fuel blending	Containers are unpacked from drums. Contents are emptied into drums or into tanks for temporary storage. If required, pretreatment (i.e., neutralization of acids) is performed in drums or glass-lined treatment vessels. Liquids are pumped into above-ground storage tanks. Material is sent for incineration for use as a secondary fuel.
Hazardous organic solids and nonfuel organic liquids	Containers are unpacked from drums. Material is either bulked with similar items for incineration as a bulk waste stream or sent to a rotary kiln incinerator for disposal in lab pack configuration.
Hazardous inorganic liquids (solutions) and solids	Containers are removed from drums. Compatible corrosive acids and bases are bulked (separately) into treatment tanks and neutralized. Oxidation/reduction reactions are performed in treatment vessels to destroy inorganic oxidizers and reducing agents. Compatible inorganic metal salts are bulked in 55-gallon drums for disposal as bulk waste streams, thereby minimizing the number of drums being disposed of in a hazardous landfill.
Nonhazardous waste	Containers are removed from drums. Items are bulked into 55-gallon drums for disposal as bulk waste streams. Options for landfilling or incineration of the nonhazardous waste is provided to the generator. In either case, the waste is handled as a bulk waste, which minimizes the number of drums (and cost) for disposal.

Chlorinated solvents and materials for recycle	Containers are removed from drums. Similar materials are bulked into larger containers for temporary storage. When a sufficient quantity of a particular material is obtained, drums (or tankers) are transported to a reputable recycler and the material is reclaimed for resale as a product.

12.4 CONCLUSION

The chemical redistribution and disposal project was completed statewide with 97% of Iowa's 436 school districts participating. During the Phase I cleanup, approximately 28000 containers were transported from 900 buildings, repacked by SET, and disposed of according to state and federal guidelines. The cost for the operation averaged $575 per district, or approximately $7.25 per pound. Those costs would actually be lower today since SET now has its own redistribution and incineration site in Texas. The most impressive statistic is that this entire process was conducted without a single incident or injury.

The plan was successful due to four primary factors: first, the extensive cooperation among governmental agencies; second, all activities were based on a "safety first" mentality; third, SET had the experience and ability to perform the pickup and disposal according to the schedule and prices they stated in writing; and fourth, the entire process was standardized and automated due to the availability of the computerized safety program.

BIBLIOGRAPHY

Code of Federal Regulations, 40, Part 261, Environmental Protection Agency, Washington, DC, July 1985.

J. Gerlovich, T. Gerard, and K. Hartman, *The Total Science Safety System,* Jakel, Inc., Des Moines, IA, 1990, Sargent-Welch Scientific Co., Skokie, IL, 1990.

Memo from J. A. Gerlovich, Iowa Department of Education to Area Education Agency Science Consultants, entitled, "Don't Let Hazardous Chemicals Blow You Away," December 30, 1988.

State of Iowa (Sep. 21, 1988). "Right-to-Know Rules," Iowa Division of Labor.

APPENDIX 1

PLANNING FOR PURCHASING FOR CHEMICAL STORAGE

JOHN BEQUETTE
Scientific Materials Management Consultant
Columbia, Missouri

Each year millions of dollars are spent to purchase chemicals for research, education, and manufacturing. Of these, many dollars' worth of chemicals are wasted due to poor planning, improper storage, and lack of safety considerations.

The key to minimizing the amount of wasted dollars is good planning based on accurate information, such as previous usage rates, projected future usage rates, available storage space for different types of chemicals, safety regulations concerning storage of hazardous chemicals, ability of personnel to handle certain size containers of different chemicals, the economic feasibility of purchasing in large quantities, and the method and time required to replenish in-house stocks.

Planning should begin with management or administration defining clear-cut areas of responsibilities and purchasing procedures. Strict enforcement in these areas will eliminate delays in ordering and deliveries, duplicate shipments, wrong addresses on shipments, and confusion in accounting areas. All of these can be equated into lost dollars and cents by the work-hours and expense incurred to correct errors.

Each of the areas needed for proper planning must be considered alone and in conjunction with all other items, starting with previous usage rates obtained from records of past purchases and consumption rates. Accurate records are a necessity for accurate planning. (Computer systems covered in Chapters 7 and 9 can be an excellent tool in the storage of this information, which can be made readily available to management and purchasing.)

Projected usage may be obtained from investigators or from increases in number of staff using chemicals or common solvents. Changes in procedures in industry or research or in projected enrollment can also be used to project needs.

Available storage space is a very important item in the planning for purchasing because the shifting or relocating of items from normal storage areas may create a burden on buildings and personnel. Also, without good records of locations, items may be forgotten until they become old and unusable and create a hazard. The compatibility of items in storage areas must also be taken into consideration.

Federal, state, and institutional safety regulations concerning the storing and handling of different types of chemicals must be taken into consideration prior to purchasing to prevent violations. The type of violation most often found is flammables in both laboratories and storage areas.

The ability of personnel to handle certain size containers and the economic feasibility of purchasing items in large quantities are two areas so closely related that they are difficult to separate. It is not economical to purchase large containers of items like mineral acids if they cannot be transferred without endangering the health and safety of personnel. Nor is it wise to purchase large containers of chemicals that are known to become dangerous as a result of peroxide crystal formation. These create a hazard to existing structures as well as personnel. Anyone who has had to dispose of a drum of ethyl ether is well aware of this fact.

Purchasing in large containers will often dictate the additional purchase of containers for transfer, plus other materials such as pumps, funnels, and related safety items. Work-hour costs and the costs of necessary equipment may eliminate any savings derived from purchasing in large quantities. This method also denies the end user the pertinent information contained on the manufacturer's label, such as handling instructions and first aid information. Also, a large container may be removed from the stock room and located in one laboratory, necessitating additional purchases of the same chemical for other laboratories. A good example of this is bromine, purchased in a 6-lb. bottle and taken to a laboratory that needs only 2 ounces.

There are many different methods of ordering chemicals, and your own procedure will indicate to you which is best for your particular operation. The method most preferred is a rapid order and delivery system, which often requires the use of open orders or blanket orders to designated suppliers. This allows the purchasing agent to place an order without creating a new purchase order or getting administrative approval for each order, thereby saving time and money.

The advantage of this method is that it allows you to utilize the supplier's storage area until the item is needed at your location. In addition, you receive newer stock on items that tend to deteriorate on the shelf, and your inventory is maintained at a lower level, reducing the amount of investment and the amount of hazardous material on hand in case of an accident. Depending on your location, the materials can be delivered on the same day or within a few days.

The disadvantage is that you will seldom receive the maximum discount that you would receive from buying in case lots or large shipments, and often you must pay freight on small shipments. Also, control of expenditures is transferred from management to purchasing, sometimes creating a situation where unauthorized purchasing can exist.

Ordering to replace items as used, on a daily or weekly basis, requires a system that monitors receiving, issues, on order, and on backorder. In most cases for this type of system to function, stock levels must be set and adhered to. The disadvantage of this type of purchasing is that it requires more work-hours than other systems to make it operate properly, and any changes in usage can affect the set stock levels. On the other hand, for large-volume users, this method allows the purchaser to allocate less storage space for items and maintain a smaller inventory.

Another method preferred by large users is a contract method. Yearly requirements are contracted for at a certain price, to be shipped at designated times during a given period or when requested. This again allows users to use storage area of the suppliers, but it also commits users to receive all materials for which they contract, even if these chemicals are no longer needed at that facility.

A third method is an annual or semiannual order. This will normally receive the best discount available but requires the most planning, taking into

TABLE A1.1   Purchasing Checklist

Item to be ordered	_____	
Amount used past _____ months	_____	
Amount used past _____ days	_____	
Projected requirements	_____	
Based on	_____	
Proper storage space available	Yes _____	No _____
Description	_____	
Equipment on hand to dispense safely	Yes _____	No _____
(If no, use additional page to list equipment needed and cost)		
Qualified personnel to dispense	Yes _____	No _____
Price per _____ in small quantity	_____	
Price per _____ in bulk quantity	_____	
Amount of savings	_____	
Other advantages, if any	_____	
Recommended purchasing procedure _____		
Recommendations and approval _____		
Recommendations and approval _____		
Recommendations and approval _____		
Action taken _____	Date _____	

consideration the following seven items: (a) previous usage, (b) projected usage, (c) available storage space, (d) safety regulations, (e) ability of personnel involved, (f) economics of bulk buying, and (g) alternative purchasing procedures. A combination of any of these methods may work well for some users, depending on location and the volume of chemicals and supplies used (Table A1.1).

Planning for purchasing is essential to the operation of any facility that uses chemicals as a result of the increased cost of chemicals, the necessity of safety in the storage and handling of chemicals, and the ever-increasing regulations and cost of disposal of these items. So, with good data, good storage, trained individuals, and good planning, chemical consuming facilities may be operated wisely and safely.

APPENDIX 2

CHEMICAL STORAGE CHECKLIST

Taking stock of current storage conditions and procedures is the first step in managing a safe chemical storeroom. The following checklist* has been developed to help assess safety in the storeroom. The checklist format not only facilitates a systematic assessment of storage and housekeeping conditions, but also identifies general and specific areas of concern. The completed checklist serves as a record of needed improvements. An affirmative answer to each item indicates a satisfactory storage condition.

Yes   No   NA†

STORAGE AREAS

Storage rooms are properly marked or identified.

Storage areas are secured whenever not in use and are available only to authorized personnel.

Storage areas are free of blind alleys.

Storage areas have two or more clearly marked exits.

Storage areas are well illuminated.

Storage areas are well ventilated, with exhaust air leaving the building. (Beware of recirculating systems.)

Storage areas have adequate air-conditioning and/or dehumidifier systems to provide a cool, dry atmosphere.

Open flames, smoking, and localized heating units are not permitted in chemical storage areas.

Mixing or transfer of chemicals is not allowed in storage areas.

*The Chemical Storage Checklist originally appeared as part of an article: "Safe Storage of Chemicals: A Checklist for Teachers," *The Science Teacher,* **48**, (2) (February 1981). Reprinted with permission from *The Science Teacher* (February 1981), published by the National Science Teachers Association, Washington, DC, and Lab Safety Supply Company, Janesville, WI.
† Not applicable.

Yes   No   NA

Aisles in the storage area are free from obstruction.

Ladders with handrails are available where needed.

Sources of sparks are completely eliminated from the storage area.

SHELF STORAGE

Large bottles and containers are stored on shelves no higher than 2 feet from the floor.

Containers of chemicals are stored below eye level.

Shelves have raised edges or rim guards to prevent the accidental dislodging of containers.

Reagent bottles or containers do not protrude over the shelf edges.

Enough space is available so that chemicals are not overcrowded.

Empty bottles are removed from stockroom shelves.

Shelves are level and stable. Shelving units are securely fastened to wall or floor.

Weight limit of shelves is posted and not exceeded.

Shelves are clean—free of dust and chemical contamination.

STORAGE CONTAINERS

Storage containers are inspected periodically for rust, corrosion, or leakage.

Damaged containers are removed or repaired immediately.

Chemicals are kept in airtight bottles, not in beakers or open vessels.

Stoppers form an airtight seal with containers.

Stoppers are easily removed from bottles or containers.

Containers of mercury are well stoppered.

Carboys are used for storing chemical solutions.

Eye-dropper bottles are not used for storing corrosive or water-reactive chemicals.

Yes   No   NA

All carboy spigots are leaktight and drip free.

Dispensing tubes on carboys are free of corrosion or aging.

LABELING OF CHEMICAL CONTAINERS

National Fire Protection Association (NFPA) hazard labels and identification system are used for labeling all dangerous chemicals.

All containers are clearly labeled as to contents.

Labels are readable and free of encrustation or contamination.

Labels are firmly attached to containers.

Chemical containers are labeled with the appropriate warning (e.g., poison, corrosive, etc.).

All container labels include both date of receipt and anticipated disposal.

Labels include precautionary measures for the specific chemical.

HOUSEKEEPING

Cleanliness and order are maintained in the storage areas at all times.

Unlabeled, contaminated, or undesirable chemicals are discarded properly.

Chemicals in storage cabinets and on shelves are inspected for decomposition on a regular basis. An inspection log is kept.

Unused chemicals are never returned to stock bottles.

Packing materials and empty cartons are removed at once from the stockroom.

Waste receptacles are properly marked and easily located.

Separate disposal containers are available for broken glass.

Environmentally safe disposal methods have been arranged for hazardous waste chemicals.

Yes   No   NA
_____

GAS CYLINDERS

All gas cylinders are secured to prevent falling over.

Gas cylinders are stored away from direct or localized heat, open flames, or sparks.

Gas cylinders are stored in a cool, dry place away from corrosive fumes or chemicals.

Gas cylinders are stored away from highly flammable substances.

Empty gas cylinders are labeled EMPTY or MT.

Empty gas cylinders are stored separate from full gas cylinders.

Flammable or toxic gases are stored at or above ground level, never in basements.

Cylinders of incompatible gases are segregated by distance.

When gas cylinders are not in use, the valve cap is securely in place to protect the valve stem and valve.

A hand truck is available for transporting gas cylinders to and from the storage area.

FIRST AID

First aid supplies are readily available and have been approved by a consulting physician.

First aid cabinets are clearly labeled.

Emergency room staff, with medical personnel specifically trained in response to chemical exposure, are readily available.

Blankets are available for shock cases and for protection of the injured.

Supervisors are trained in CPR.

Emergency telephone numbers are posted on or near the telephone.

Eyewash and shower facilities are within 10 seconds or 100 feet of the site of hazardous materials.

Eyewash and shower facilities are periodically inspected and maintained.

Hand-washing facilities are readily available for stockroom personnel.

Yes　　No　　NA

EMERGENCY PREPAREDNESS

An emergency warning system is available in the event of an accident.

Emergency and evacuation procedures are known by stockroom personnel.

Emergency and personnel protective equipment is located *outside* areas where accidents may occur.

At least two self-contained breathing apparatuses are available for emergencies. Personnel are trained in usage.

Equipment and supplies for cleaning up spills are readily available.

Fire extinguishers are immediately accessible.

Fire extinguishers are periodically inspected and maintained.

Fire and smoke alarms are located in fire-prone areas, with periodic maintenance and inspection.

The organization has an Emergency Response Plan.

CHEMICAL STORAGE

Chemicals are not exposed to direct sunlight or localized heat.

Containers of corrosive chemicals are stored in trays large enough to contain spillage or leakage.

Chemicals are stored by reactive class (e.g., flammables with flammables, oxidizers with oxidizers).

An incompatibility-compatibility guide is available to indicate arrangement of chemicals.

Incompatible chemicals are physically segregated from each other during storage.

Acids

Large bottles of acids are stored on a low shelf or in acid cabinets.

Oxidizing acids are segregated from organic acids and flammable and combustible materials.

Acids are separated from caustics and from active metals such as sodium, magnesium, and potassium.

Yes   No   NA

Acids are segregated from chemicals that can generate toxic gases on contact, such as sodium cyanide and iron sulfide.

Bottle carriers are used for transporting acid bottles.

Absorbents or acid neutralizers are available for acid spills.

Caustics

Caustics are stored away from acids.

Solutions of inorganic hydroxides are stored in polyethylene containers.

Absorbents or caustic neutralizers are available for spills.

Flammables

Stockroom personnel are aware of the hazards associated with flammable materials.

All flammable liquid containers are in compliance with the maximum container sizes found in Table H-12, 29 CFR 1910.106.

OSHA/NFPA-specified safety cabinets are used for the storage of flammable liquids.

Flammable liquids are stored in accordance with NFPA Standard No. 30, Flammable and Combustible Liquids Code.

Flammables are kept away from any source of ignition: flames, heat, or sparks.

Approved refrigerators are used for storing highly volatile flammable liquids.

All electrical service equipment is explosion-proof for the appropriate class and group of flammable liquids.

Bonding and grounding wires are used where flammables are stored and dispensed.

Absorbents are available for leaks or spills.

Peroxide-Forming Chemicals

Peroxide-forming chemicals are stored in airtight containers in a dark, cool, and dry place.

Yes   No   NA

Peroxide-forming chemicals are properly disposed of before the date of expected peroxide formation.

Suspicion of peroxide contamination is immediately evaluated by use of safe procedures.

Chemicals are labeled with date received, date opened, and disposal date.

Water-Reactive Chemicals

Chemicals are kept in a cool and dry place.

In case of fire, a Class D fire extinguisher is used.

Oxidizers

Oxidizers are stored away from flammable, combustible, and reducing agents (e.g., zinc, alkaline metals).

Toxic Compounds

Toxic compounds are stored according to the nature of the chemical, with appropriate security employed where necessary.

A Poison Control Network telephone number is posted.

APPENDIX 3

GLOSSARY OF WORD PROCESSING AND MICROCOMPUTING TERMS

ALLEN G. MACENSKI
Operations Health and Safety Manager
TRW, Inc.
Redondo Beach, California

Access time The time required to fetch a word from memory.

Accumulator An 8-, 16-, or 32-bit register on the CPU chip that serves as workspace for arithmetic, logical, and input/output (I/O) operations. Data fetched from memory can be added, compared, tested, or otherwise operated on and the result held in the accumulator. Programmed transfers of data also pass through it.

Ada A large-scale computer language commissioned by the Department of Defense, Ada is expected to be the workhorse language of the next two decades. Like COBOL, Ada generally uses full English words; like Pascal, it is designed for modular program construction. Ada is named after Lord Byron's daughter, a pioneer computer programmer.

Address A binary number, or symbol for it, that identifies a register, cell of storage, or some other data source or destination. Eight-bit microprocessors usually use 16-bit addresses. Since there are 2^{16} or 65 536 different possible combinations of 16 bits, this allows the direct addressing of 64K (64000) locations. Sixteen-bit microprocessors use 20-bit (Intel 8086) or 24-bit (Motorola M6800) addresses that can refer to 1-million and 16-million memory locations, respectively.

Algorithm A step-by-step process for the resolution of a problem. Usually developed in outline or as a flowchart before coding, that is, setting it into computer language.

Alphanumeric The set of all alphabetic and numeric characters, along with related symbols.

Analog Refers to data in the form of continuously changing physical quantities—waves—or devices that operate on it.

APL   A high-level language pioneered by Kenneth Iverson that uses unique keyboard symbols and specially developed mathematical operators.

Arithmetic logic unit (ALU)   The element in a computer that can perform the basic data manipulations in the central processor.

Array computer   A computer in which the microprocessors are wired together into an "array" of so many rows and columns. The array computer is extremely fast for two reasons: (a) it can process several commands at the same time, and (b) it can begin processing new information at the same time that it is processing old information.

Artificial intelligence (AI)   Approximation by a computer and its software of certain functions of human intelligence, learning, adapting, reasoning, self-correcting, and automatic improvement. It can change its own program to adapt to new data it encounters. It is expected to be an important component of the fifth-generation computers of the 1990s.

ASCII   American standard code for information interchange. This is a character code used for representing information.

Assembler   Program that takes the mnemonic form of the assembly language and converts it into binary object code for execution.

Assembly language   Developed in the early 1950s so that programmers could use abbreviated word commands (mnemonics) in place of the confusing strings of ones and zeros that are the direct representation of a computer's language.

Bandwidth   The maximum number of data units that can be transferred along a channel per second.

Batch   A processing mode whereby a program is submitted to the computer and the results are delivered back. No interactive communication between program and user is possible.

Baud   The rate at which data are transmitted over a serial link, such as a telephone line, in bits per second. The format for data transmission is 10 or 11 bits per character, so 300 baud is about 30 characters per second.

Bit   The contraction of "binary digit." A bit always has the value zero or one. Bits are universally used in electronic systems to encode information, orders (instructions), and data. Bits are usually grouped in nybbles (4), bytes (8), or larger units.

Bootstrap   A program used for starting the computer, usually by clearing the memory, setting up devices, and loading the operating system from input/output internal or external memory.

Bubble memory   A memory device placed on a chip. The bubbles are microscopically small, magnetized "domains" that can be moved across a thin magnetic film by a magnetic field. Magnetic-bubble chips that store a million bits of information have been fabricated. Magnetic bubble memory is not as fast as RAM or ROM, but many times faster than mass memory devices such as tapes and disks. Yet like tapes and disks, the information stored inside the magnetized bubbles is retained even after the computer's power is switched off.

Byte A group of 8 bits treated as a unit. Under the ASCII format, 7 of the bits provide 128 different codes representing characters, and the eighth is a paritybit for error checking.

CCD Charge-coupled device. A mass storage device for information on a single chip.

Coding Putting an algorithm into computer language.

Compiler A translation program that converts high-level instructions into a set of binary instructions for execution. Each high-level language requires a compiler or an interpreter. A compiler translates the complete program, which is then executed. Every change in the program requires a complete recompilation.

CPU Central processing unit. The computer module in charge of fetching, decoding, and executing instructions.

Data security Protection of computerized information by various means, including cryptography, locks, identification cards and badges, restricted access to the computer, passwords, physical and electronic backup copies of the data, and so on.

Diagnostic A program or routine used to diagnose system malfunctions.

Digital Refers to data in the form of discrete units—"on/off" or "high/low" states—and to devices operating on such data.

DIP Dual in-line package. A standard integrated circuit (IC) package with two rows of pins at 0.1-inch interval.

Disable To render a control or process temporarily inoperable while another process takes place.

Disk drive The machinery that contains, rotates, writes onto, and reads from a disk.

Documentation Instructions and other explanatory materials supporting computer hardware and software.

Dump To copy the contents of memory or disk to video display, printer, or storage device so that it can be checked out in detail or kept as backup.

Expansion interface A device to expand the functional capacity of a computer by containing additional memory or controlling more peripherals.

External memory Used to store programs and information that would otherwise be lost if the computer was turned off. Cassette tapes, disks, bubble memory, and CCDs (charge-coupled devices) are also known as *mass memory* and *removable memory.*

File A logical block of information, designated by name, and considered as a unit by a user. A file may be physically divided into smaller records.

Firmware A program permanently fixed onto a memory chip (ROM), that is, software on a hardware support.

Flat panel displays   Thin, light, with low-power consumption, free from geometric distortion, these displays will begin replacing CRTs in some microcomputers this year. Most familiar is the passive (non-light-emitting) LCD (liquid crystal display). Active (light-emitting) technologies include gas plasma and thin-film electroluminescence (EL). Electroluminescence seems the most likely to develop into a replacement for the CRT in terms of response time, gray scale, and resolution.

Flow chart   Symbolic representation of a program sequence. Boxes represent orders of computations. Diamonds represent tests and decisions (branches). A flow chart is the recommended step between algorithm specification and program writing. it greatly facilitates understanding and debugging by breaking down the program into logical, sequential modules.

FORTH   A very efficient, high-level, interpreted language especially suited to microcomputers used in conjunction with sensors and control systems. FORTH's basic commands, very close to assembly language, can be used to define new procedures made by the user. It is called a "threaded language" because these new procedures are created by threading together old ones.

Grid   A division of the computer screen into evenly spaced horizontal and vertical lines. Used for locating points on the screen. These points are expressed as row and column coordinates.

Hashing   A system for verifying data input.

Heuristic methods   Methods that serve to guide or reveal answers that cannot be proved.

High-level languages (HLL)   Problem-oriented languages, much easier to use than machine-oriented languages. They are faster to write than assembly language but produce less efficient object code. COBOL, FORTRAN, PL/1, RPG, and Ada are high-level languages usually compiled. BASIC, Pascal, LISP, and FORTH are usually interpreted. One high-level language statement can produce 10 machine-language statements.

I/O   The communication of information to and from a computer or peripheral device.

Input device   Any machine that allows commands or information to be entered into the computer's main (RAM) memory. An input device could be a typewriter keyboard, an organ keyboard, a tape drive, a disk drive, a microphone, a light pen, a digitizer, or an electronic sensor.

Integrated circuit (IC)   A complex, microscopic circuit on a chip of silicon.

Interface   The hardware or software required to interconnect a device to a system. One-chip interfaces now exist for most peripherals.

ISAM   Indexed sequential access method. A method for organizing data files for rapid access.

Kilobyte   *Kilo* means 1000, but *kilobyte* means, precisely, 1024 bytes.

Language   In relation to computers, any unified, related set of commands or instructions that the computer can accept. Low-level languages are difficult to use but closely resemble the fundamental operations of the computer. High-level languages resemble English.

Light pen An input device for CRT. It records the emission of light at the point of contact with the screen. The timing relationship to the beginning of a scan tells the computer its position on the screen.

LISP List processing language. A high-level, interpreted language especially effective at recursion and the manipulation of symbolic strings. Good for writing languages, including itself, and for the development of software for artificial intelligence.

Load The action of transferring data in a register or memory location, or a program into a memory area.

Location The physical place in the computer's memory, with a unique address, where an item of information is stored.

Logic The term used to designate that part of the computer's circuitry that makes logical decisions.

Logical decision The capability of the computer to decide whether one quantity is greater than, equal to, or less than another quantity and then use the outcome of that decision as a cue to proceed in a given way with a program.

Logical operator Symbols used in programming to represent the operations of logic, including AND, OR, and NOT. Expressions containing these symbols are often called *Boolean expressions* (after George Boole, a nineteenth-century British mathematician and logician).

LOGO A computer language, developed by scientists at MIT's Artificial Intelligence Laboratory, that has been used experimentally over the past several years in a children's learning laboratory as a tool to help youngsters master concepts in mathematics, art, and science.

Loop A group of instructions that may be executed recursively.

LSI Large-scale integration.

Machine language Set of binary codes, representing the instructions that can be directly executed by a processor

Mainframe The box that houses the computer's main memory and logic components—its CPU, RAM, ROM, I/O interface circuitry, and so on. The word is also used to distinguish the very large computer from the minicomputer or microcomputer; it usually uses 32-bit term.

Main memory The internal memory of the computer contained in its circuitry, as opposed to peripheral memory (tapes, disks).

Maintenance The adjustment of an existing program to allow acceptance of new tasks or conditions (e.g., a new category of payroll deduction).

Mask A pattern, usually "printed" on glass, used to define areas of the chip on the wafer for production purposes.

Mass memory See external memory.

Menu A list of programs or applications that are available by making a selection. For example, a small home computer might display the following menu: Do you want to (a) balance checkbook, (b) see appointments for May, or (c) see a recipe? Type letter desired.

Merge A computerized process whereby two or more files are brought together by a common attribute, as zip codes in ascending numerical order.

Microcomputer A small but complete computer system, including CPU, memory, I/O interfaces, and power supply. Generally uses 8-bit word.

Microprocessor Large-scale integration implementation of a complete processor (ALU + control unit) on a single chip.

Minicomputer A small computer, intermediate in size between a microcomputer and a large computer. Uses 15-bit word.

Model A computer reproduction (or simulation) of a real or imaginary person, process, place, or thing. Models can be simple or complex; artistic, educational, or entertaining; serious or part of a game.

Modem Doculator–demodulator. A device that transforms a computer's electrical pulses into audible tones for transmission over a telephone line to another computer. A modem also receives incoming tones and transforms them into binary impulses that can be processed and stored by the computer.

Multiplexing Transmitting several signals simultaneously over one data channel in a communications system. Some microprocessors multiplex addresses and data to memory; others use separate address and data channels.

Multiprocessing A small computer that has more than one CPU and is thus able to process several instructions simultaneously; another form of parallel processing.

Nanosecond A billionth of a second. Most computers have a cycle time, or "heartbeat," of hundreds of nanoseconds. Most high-speed computers have a cycle time of around 50 nanoseconds.

Natural language A spoken, human language such as English, Spanish, Arabic, or Chinese. In the future, small computers will probably be fast enough and have large enough vocabularies to enable one to talk to them, using one's native language. Yet there will still be a need for special computer languages that are more efficient than native languages for handling certain kinds of tasks.

Network Several intelligent devices such as microcomputers interconnected so that they can send instructions and data back and forth between themselves, thus forming a larger computational system.

Nybble Usually 4 bits, or half of a byte.

Object program A program in machine-readable form. A compiler translates a source program into an object program.

OCR Optical character recognition. The data input system that allows the computer to "read" printed information.

On-line   Directly connected to the computer system and in performance-ready condition.

Operating system (OS)   Software required to manage the hardware and logical resources of a system, including scheduling and file management.

Optical wand   An input device with a photoelectric camera that senses black and white light patterns.

Output device   A machine that transfers programs or information from the computer to some other medium. Examples of output devices include tape disk and bubble memory-drives; computer printers, typewriters, and plotters; the computer picture screen (video monitor); robots; and sound synthesis devices that enable the computer to talk and play music.

Pascal   A high-level language developed to teach structured programming. Pascal has become popular for general use.

PCB   Printed circuit board.

Peripheral   Any human interface device connected to a computer.

Picosecond   A trillionth of a second. Even light, the fastest substance in the universe, can travel only one one-hundredth of an inch in a trillionth of a second.

PILOT   An interpreter language, like BASIC and FORTH. PILOT is a simple programming language that is ideally suited for use in computer-aided instruction applications.

Pipeline computer   A computer with a string of "processor" CPUs all capable of executing an operation simultaneously.

Pixel   The computer picture screen is divided into rows and columns of tiny dots, squares, or cells. Each of these is a pixel (from picture element).

PL/I   A high-level, compiled language developed by IBM during the 1960s that combines the best features of COBOL and FORTRAN and generates efficient object code. Its large vocabulary and large compiler were only recently adapted for 8-bit microcomputers.

Plasma-ray device   A flat computer picture screen based on a gird of metallic conductors separated by a thin layer of gas. When a signal is generated at any intersection along the grid, the gas discharges and causes the transparent screen to glow at this point. In the future, the thinner, more reliable, plasma ray and solidstate devices will replace CRTs as computer video display terminals.

Plotter   A mechanical device for drawing lines under computer control.

Portability   The property of software that permits its use in a variety of computer environments.

Program   A sequence of instructions that results in the execution of an algorithm. Programs are essentially written at three levels: (a) binary (can be directly executed by the MPU), (b) assembly language (symbolic representation of the binary); and (c) high-level language (e.g., BASIC), requiring a compiler or an interpreter.

RAM Random access memory. Denotes in fact addressable Read/Write LSI memory.

Random access An access method whereby each word can be retrieved directly by its address.

Read To accept data from a disk, card, and so on, for storage and/or processing.

Real time Immediate and concurrent processing.

Record The process of writing information, or the block of information itself.

Register One-word memory, usually implemented in fast flip-flops, directly accessible to a processor. The CPU includes a set of integral registers that can be accessed much faster than the main memory.

Resolution The quality of the image on the CRT, as influenced by the number of pixels on the screen. The greater the number of pixels, the higher the resolution.

Robot Any stored-program device capable of altering its external environment.

ROM Read-only memory.

Sensor Any device that acts as "eyes" or "ears" for a small computer. Types of sensors include photoelectric sensors that are sensitive to light, image sensor cameras that record visual images and transform the images into digital signals, pressure sensors that are sensitive to any kind of pressure, contact sensors that record IR information, and ultrasonic transducers that produce a high-frequency sound wave that bounces off objects and lets the computer calculate the distance between itself and those objects.

Sequential access The method in which data are accessed by scanning blocks or records sequentially.

Service bureau A company that offers time-sharing, programming, and other computer services to businesses.

Simulation A computerized reproduction, image, or replica of a situation or set of conditions.

Smalltalk A computer language developed by the Xerox Palo Alto Research Center. When it finally becomes commercially available, Smalltalk may prove to be the most powerful language for small computers of the future.

Soft copy Information contained magnetically in storage.

Sort A full or part program to reorder data sequentially, usually in alphabetic or numeric order.

Source program A program written in a language that is not directly readable by a computer; it must be converted to an object program for use by a computer.

Static RAM Unlike ordinary, volatile memory, static memory retains its contents even when the main current is turned off. The trickle of electricity from a battery is sufficient to refresh it.

String A linear series of symbols treated as a unit, such as this sentence.

String floppy An endless loop of recording tape in a cartridge used as external memory.

Structured programming In a program, proceeding in a systematic way from section to section rather than branching widely on GOTO in instructions.

Subroutine A programmed module for a special task (e.g., computing a square root) that can be called in at any point of the main program.

Tape Inexpensive mass storage medium. Must be accessed sequentially. Convenient for large files.

Telepresence By using all of a robot's sensors, a person can be electronically aware of the robot's immediate environment and control the robot's actions just as if the person were actually in the location of the robot.

Telescreen A two-way, audiovisual television used to monitor and control remote activities.

Teletext Textual information transmitted to people's homes through television. Information is usually maintained and updated on a computer. A teletext system often allows two-way interaction by creating a viewer–computer link over a telephone line.

Throughput The number of instructions executed or the amount of data transmitted per second. A measure of a computer's power and efficiency; 10 MIPS means 10 million instructions per second.

Timesharing Occurs when a single computer has multiple users who are each getting a "slice" of each second of the computer's processing time.

Transportable Of software: usable on many different computers.

Utility Software for routine tasks or for assisting programmers.

Videodisk A recordlike device storing a large amount of audio and visual information that can be linked to a computer. A single side of one videodisk can store the pictures and sound for 54 000 separate television screens. Yet any one of these screens' images can be accessed and displayed.

Virtual memory A major extension of main memory address space into a secondary memory such as hard disk. A program in virtual memory is divided into segments called *pages,* and these pages are read into main memory as needed.

VLSI Very large scale integration. In practice, the compression of more than 10 000 transistors on a single chip.

Wafer Three-or four-inch slice of silicon that is overlaid with microscopic circuitry and broken up into many individual computer chips.

Word The unit of information stored in a computer's memory, moved in parallel along its data paths, and worked on in its registers. Word size is an important distinction between classes of computer. Small microprocessors have a 4-bit (one nybble, or half-byte) word. The classic microcomputer has an i-bit (one byte) word, although micros with 16-bit words are now appearing. Minicomputers usually use 16-bit words. Mainframes use words of 32 bits or more in some cases.

Write   To transfer data from internal to external memory.

APPENDIX 4

ON-LINE DATABASE SOURCES: CHEMICAL HEALTH AND SAFETY

ALLEN G. MACENSKI
Operations Health and Safety Manager
TRW, Inc.
Redondo Beach, California

A4.1 NONBIBLIOGRAPHIC FACTUAL DATABASES (HEALTH AND TOXICOLOGY DATA SOURCES)

Hazardous Substances Data Bank (HSDB): 4100+ chemicals
Detailed, reviewed chemical profiles including 144 data fields in 10 categories. Fully referenced and partially peer reviewed (5).

Hazardline: 4000+ chemicals
Emergency response, safety, regulatory, and health information profiles (6).

Material Safety Data Sheets (MSDSs): 3200+ chemicals
Information extracted from Hazardline profiles in MSDS format (6).

Oil and Hazardous Materials Technical Assistance Data System (OHMTADS): 1400+ chemicals
Limited toxdata; emphasis on environmental and safety data for spills response with 126 fields (2,4).

Chemical Hazard Information Response System (CHRIS): 1016 substances
Produced by the U.S. Coast Guard; provides information for emergency response during transport of hazardous chemicals. Contains information on labeling, physical and chemical properties, fire hazards, chemical reactivity, water pollution, and hazard classifications (2).

Chemical Evaluation Search and Retrieval System (CEASERS): 194+ chemicals
Detailed information and evaluations on chemicals of particular importance in the great Lakes Basin. Each record has 185 data fields (2,4).

Chemical Carcinogenesis Research Information System (CCRIS): 1200+ chemicals

Contains individual evaluated assay results and test conditions for substances in the areas of carcinogenicity, mutagenicity, tumor production, and cocarcinogenicity (2,4,5).

Toxicological Profiles, ATSDR and EPA

(Available November 1987 on Toxnet.)

DIRLINE

Contains information on reference centers, information resources, and individuals or groups who are willing to provide users with information (7).

National Pesticide Information System (NPIRS): 110 active ingredients

EPA Fact Sheet Database, Office of Pesticide Program newsletters, federally/state registered products (13).

Pesticide Databank, PESTMAN: 5000+ compounds

On-line version of *Pesticide Manual.* Profiles produced by British Crop Protection Council in collaboration with Commonwealth Agricultural Bureau (9).

A4.2 BIBLIOGRAPHIC DATABASES (CITATIONS TO JOURNAL ARTICLES AND RESEARCH PAPERS)

National Technical Information Service (NTIS): 1965 to present; 1 200 000+ citations

Reports of U.S. government-sponsored research and development (1,3,10,11).

Life Sciences Collection: 1978 to present; 847 000+ citations

Section corresponding to *Toxicology Abstracts* (3).

HSELine: 1977 to present; 59 000+ citations

Covers all Health and Safety Commission and Health and Safety Executive (British OSHA) publications and miscellaneous health and safety documents (7,8).

Agricola: 1970 to present; 2 229 000+ citations

Database of the National Agricultural Library. Covers pesticides, fertilizers, and other agriculturally related subjects (3).

Cancerlit: 1963 to present; 520 000+ citations

Abstracts from *Carcinogenesis Abstracts, Cancer Therapy Abstracts,* and other sources (3,7).

Chemical Exposure: 1984 to present; 16 000+ violations
Body-burden information and references (3).

CIS—ILO: 1976 to present
Indexes worldwide literature on occupational safety and health (7,9).

Pre-Med: Current 4 months; 5000+ citations
Current clinical medicine from 108 core medical journals (1).

DATA SYSTEMS REFERENCED

1. BRS, Bibliographical Retrieval Services, 1900 Rt. 7, Latham, NY; 800-833-4707.
2. CIS, Chemical Information Systems, Inc., 7215 York Road, Baltimore, MD 21212; 301-821-5980.
3. Dialog, Dialog Information Systems, Inc., 3460 Hillview Avenue, Palo Alto, CA 94304; 800-3-DIALOG.
4. ICIS, Out of business.
5. NLM, National Library of Medicine, Specialized Information Services, Building 38A, 8600 Rockville Pike, Bethesda, MD 20209; 301-496-1131.
6. OHS, Occupational Health Services, Inc., 400 Plaza Drive, Secaucus, NJ 07094; 201-865-7500.
7. OSH-ROM, Silver Platter Information, Inc., 37 Walnut Street, Wellesley Hills, MA 92181; 617-239-0306 (CD-ROM for IBM PC and compatibles).
8. Pergamon Infoline, Pergomon Infoline, Inc., 1340 Old Chain Bridge Road, McLean, VA 22010; 703-442-0900.
9. Questel, Questel, 5201 Leesburg Pike, Suite 603, Falls Church, VA 22041; 703-845-1133.
10. SDC, SDC Information Services, 22500 Colorado Avenue, Santa Monica, CA 90406; 213-820-4111 (merged with Pergamon Infoline).
11. STN, STN International, Chemical Abstracts Service, P.O. Box 2228, Columbus, OH 43202; 800-848-6538.
12. HAZMAT, The Fire Service Software Company, 134 Middle Neck Road, Suite 210, Great Neck, NY 11021; 516-829-5858.
13. NPIRS, National Pesticide Retrieval System, Entomology Hall, Purdue University, West Lafayette, IN 47907; 317-494-6614.

APPENDIX 5

FLASH POINTS OF COMMON FLAMMABLE LIQUIDS

The following are Class IA flammable liquids (flash point <73°F; boiling point <100°F):

Flammable Liquid	Flash Point (°F)
Ethyl chloride	−58
Pentane	−57
Ethyl ether	−49
Acetaldehyde	−36
Isopropylamine	−35
Ethyl formate	− 2
Ethylamine	0

The following are Class IB flammable liquids (flash point <73°F, boiling point ⩾100°F):

Flammable Liquid	Flash Point (°F)[a]
Naphtha[b]	−40 to 68
Allyl chloride	−25
Carbon disulfide	−22
Isopropyl ether	−18
Acrolein	−15
Hexane	− 7
Cyclohexane	− 4
Ethyl bromide	<− 4
Nickel carbonyl	− 4
Acetone	1.4
1,1-Dimethylhydrazine	5
Tetrahydrofuran	6
Butyl amine	10
Benzene	12
Methyl acetate	14
Methyl ethyl ketone	21

Flammable Liquid	Flash Point ($^\circ$F)[a]
Ethyl acetate	24
Heptane	25
Acrylonitrile	30
Butyl mercaptan	35
Toluene	40
2-Pentanone	45
Methyl methacrylate	50 (oc)
Methanol	52
Isopropanol	53
Dioxane	54
Ethylene dichloride	55
Octane	56
Propanol	59
sec-Butyl acetate	62
Pyridine	68
Allyl alcohol	70
Butyl acetate	72

[a] Closed cup values are given unless where denoted by "oc" (open cup).
[b] Borderline Class IA.

The following are Class IC flammable liquids (flash point $\geqslant 73\,^\circ$F, but less than $100\,^\circ$F):

Flammable Liquid	Flash Point ($^\circ$F)[a]
Methyl isobutyl ketone	73
2-Butanol	75
n-Amyl acetate	77
2-Hexanone	77
Isoamyl acetate	77
Xylene	81
Butyl alcohol	84
Chlorobenzene	84
p-Ansidine	86
sec-Amyl acetate	89
Styrene	90
Ethylene diamine	93
Morpholine	95
Turpentine	95

[a] Flash-point values were taken from *NIOSH/OSHA Pocket Guide to Chemical Hazards*, DHEW (NIOSH) Publication No. 78-210, Fourth printing, August 1981.

INDEX